STATA CROSS-SECTIONAL TIME-SERIES
REFERENCE MANUAL
RELEASE 8

A Stata Press Publication
STATA CORPORATION
College Station, Texas

Stata Press, 4905 Lakeway Drive, College Station, Texas 77845

The suggested citation for this software is

StataCorp. 2003. *Stata Statistical Software: Release 8.0*. College Station, TX: Stata Corporation.

Table of Contents

Cross-Referencing the Documentation

When reading this manual, you will find references to other Stata manuals. For example,

[U] **29 Overview of Stata estimation commands**
[R] **regress**
[P] **matrix define**

The first is a reference to Chapter 29, *Overview of Stata estimation commands* in the *Stata User's Guide*, the second is a reference to the `regress` entry in the *Base Reference Manual*, and the third is a reference to the `matrix define` entry in the *Programming Reference Manual*.

All of the manuals in the Stata Documentation have a shorthand notation, such as [U] for the *User's Guide* and [R] for the *Base Reference Manual*.

The complete list of shorthand notations and manuals is as follows:

[GSM]	*Getting Started with Stata for Macintosh*
[GSU]	*Getting Started with Stata for Unix*
[GSW]	*Getting Started with Stata for Windows*
[U]	*Stata User's Guide*
[R]	*Stata Base Reference Manual*
[G]	*Stata Graphics Reference Manual*
[P]	*Stata Programming Reference Manual*
[CL]	*Stata Cluster Analysis Reference Manual*
[XT]	*Stata Cross-Sectional Time-Series Reference Manual*
[SVY]	*Stata Survey Data Reference Manual*
[ST]	*Stata Survival Analysis & Epidemiological Tables Reference Manual*
[TS]	*Stata Time-Series Reference Manual*

Detailed information about each of these manuals may be found online at

http://www.stata-press.com/manuals/

Title

> **intro** — Introduction to cross-sectional time-series manual

Description

This entry describes the *Stata Cross-Sectional Time-Series Reference Manual.*

Remarks

This manual documents the xt commands, and is referred to as [XT] in cross-references. Following this entry, [XT] **xt** provides an overview of the xt commands.

This manual is arranged alphabetically. If you are new to Stata's xt commands, we recommend that you read the following sections first:

> [XT] **xt**   Introduction to xt commands
> [XT] **xtreg** Fixed-, between-, and random-effects, and population-averaged linear models

Stata is continually being updated. Stata users are always writing new commands, as well. To find out about the latest cross-sectional time-series features, type search panel data after installing the latest official updates; see [R] **update**.

What's new

Stata has two new cross-sectional time-series estimation commands, and various improvements to existing commands. This section is intended for previous Stata users. If you are new to Stata, you may as well skip it.

1. The new xthtaylor command fits panel-data random-effects models in which some of the covariates are correlated with the unobserved individual-level random effect. Both the Hausman–Taylor (1981) and the Amemiya–MaCurdy (1986) estimators are supplied.

   Although both the xthtaylor and xtivreg estimators use instrumental variables, these commands address different problems. The estimators implemented in xtivreg assume that the explanatory variables may be correlated with the idiosyncratic error ϵ_{it}, and the xthtaylor estimators assume that while the explanatory variables may be correlated with the individual-level random-effects u_i, the covariates are uncorrelated with ϵ_{it}.

   See [XT] **xthtaylor** for details.

2. The new xtfrontier command fits stochastic production or cost frontier models for panel data. Two different parameterizations of the inefficiency term are allowed: a time-invariant model and the Battese–Coelli (1992) parameterization of time-effects. See [XT] **xtfrontier**.

3. xtabond now allows endogenous regressors, and xtabond is now faster. See [XT] **xtabond**.

4. xtintreg and xttobit have additional predict options. Option pr0(a,b) produces the probability that the dependent variable falls within the interval (a, b); i.e., $\Pr(a < y < b)$. Option e0(a,b) produces the expected value of the dependent variable conditional on censoring; i.e., $E(y|a < y < b)$. Option ystar0(a,b) produces the expected value of the dependent variable truncated at the censoring point(s); i.e., $E(y^*)$, where $y^* = \max\{a, \min(y, b)\}$. See [XT] **xtintreg** and [XT] **xttobit**.

5. `xtregar`, `fe` now allows weights; `fweights` and `aweights` are allowed with `rhotype(regress)` or `rhotype(freg)`, or with a fixed rho; see [XT] **xtregar**.

6. `xtivreg` now saves the list of instrumented variables in `e(instd)` and the list of instruments in `e(insts)`; see [XT] **xtivreg**.

7. For `xtpcse`, the previous existing restrictions on how weights are applied have been removed; see [XT] **xtpcse**.

8. `xtgee` and `xtlogit` have a new `nodisplay` option that suppresses the header and table of coefficients. This option is for programmers. See [XT] **xtgee** and [XT] **xtlogit**.

9. Two commands have been renamed.

   `xtpois` has been renamed `xtpoisson`, but `xtpois` will still work.

   `xtclog` has been renamed `xtcloglog`, but `xtclog` will still work.

For a complete list of all new features in Stata 8, see [U] **1.3 What's new**.

References

Amemiya, T. and T. MaCurdy. 1986. Instrumental-variable estimation of an error-components model. *Econometrica* 54(4): 869–880.

Battese, G. E. and T. J. Coelli. 1992. Frontier production functions, technical efficiency and panel data: with applications to paddy farmers in India. *Journal of Productivity Analysis* 3: 153–169.

Hausman, J. A. and W. E. Taylor. 1981 Panel data and unobservable individual effects. *Econometrica* 49: 1377–1398.

Also See

Complementary:	[U] **1.3 What's new**
Background:	[R] **intro**

Title

> **xt** — Introduction to xt commands

Syntax

$xt cmd \dots \left[, \text{i}(varname_i) \ \text{t}(varname_t) \ \dots \right]$

$\text{iis} \left[varname_i\right] \left[, \text{clear} \right]$

$\text{tis} \left[varname_t\right] \left[, \text{clear} \right]$

Description

The xt series of commands provide tools for analyzing cross-sectional time-series datasets. Cross-sectional time-series (longitudinal) datasets are of the form x_{it}, where x_{it} is a vector of observations for unit i and time t. The particular commands (such as xtdes, xtsum, xtreg, etc.) are documented in the [XT] **xt** entries that follow this entry. This entry deals with concepts common across commands.

iis is related to the i() option on the other xt commands. Command iis or option i() sets the name of the variable corresponding to the unit index i. iis without an argument displays the current name of the unit variable.

tis is related to the t() option on the other xt commands. Command tis or option t() sets the name of the variable corresponding to the time index t. tis without an argument displays the current name of the time variable.

Some xt commands use time-series operators in their internal calculations, and thus require that your data be tsset. See [TS] **tsset** for details on this process. For instance, since xtabond uses time-series operators in its internal calculations, you must tsset your dataset before using it. As in [XT] **xtabond**, the manual entry for the command explicitly states that you must tsset your data before using the command. For these commands, iis and tis are neither sufficient nor recommended.

Note that specifying iis or tis will clear any previous tsset settings. Also, specifying tsset will override any settings specified by iss or tis.

If your interest is in general time-series analysis, see [U] **29.13 Models with time-series data** and the *Stata Time-Series Reference Manual*.

(Continued on next page)

Options

i(*varname$_i$*) specifies the variable name corresponding to index i in $\mathbf{x}_{it}$. This must be a single, numeric variable, although whether it takes on the values 1, 2, 3 or 1, 7, 9, or even -2, $\sqrt{2}$, π, is irrelevant. (If the identifying variable is a string, use egen's group() function to make a numeric variable from it; see [R] **egen**.)

For instance, if the cross-sectional time-series data are of persons in the years 1991–1994, each observation is a person in one of the years; there are four observations per person (assuming no missing data). *varname$_i$* is the name of the variable that uniquely identifies the persons.

You can specify the i() option the first time you estimate, or you can use the iis command to set the i() beforehand. Note that it is not necessary to specify i() if the data have been previously tsset, or if iis has been previously specified—in these cases, the group variable is taken from the previous setting. See [XT] **xt**.

t(*varname$_t$*) specifies the variable name corresponding to index t in $\mathbf{x}_{it}$. This must be a single, numeric variable, although whether it takes on the values 1, 2, 3 or 1, 7, 9, or even -2, $\sqrt{2}$, π, is irrelevant.

For instance, if the cross-sectional time-series data are of persons in the years 1991–1994, each observation is a person in one of the years; there are four observations per person (assuming no missing data). *varname$_t$* is the name of the variable recording the year.

You can specify the t() option the first time you estimate, or you can use the tis command to set the t() beforehand. Note that it is not necessary to specify t() if the data have been previously tsset, or if tis has been previously specified—in these cases, the group variable is taken from the previous setting. See [XT] **xt**.

clear removes the definition of i() or t(). For instance, typing tis, clear makes Stata forget the identity of the t() variable.

Remarks

Consider having data on n units—individuals, firms, countries, or whatever—over T time periods. The data might be income and other characteristics of n persons surveyed each of T years, or the output and costs of n firms collected over T months, or the health and behavioral characteristics of n patients collected over T years. Such cross-sectional time-series datasets are sometimes called longitudinal datasets or panels, and we write x_{it} for the value of x for unit i at time t. The xt commands assume that such datasets are stored as a sequence of observations on (i, t, x).

▷ Example

If we had data on pulmonary function (measured by forced expiratory volume or FEV) along with smoking behavior, age, sex, and height, a piece of the data might be

```
. list in 1/6, separator(0) divider
```

	pid	yr_visit	fev	age	sex	height	smokes
1.	1071	1991	1.21	25	1	69	0
2.	1071	1992	1.52	26	1	69	0
3.	1071	1993	1.32	28	1	68	0
4.	1072	1991	1.33	18	1	71	1
5.	1072	1992	1.18	20	1	71	1
6.	1072	1993	1.19	21	1	71	0

The other xt commands need to know the identities of the variables identifying patient and time. With these data, you would type

```
. iis pid
. tis yr_visit
```

Having made this declaration, you need not specify the i() and t() options on the other xt commands. If you resaved the data, you need not respecify iis and tis in future sessions.

◁

❑ Technical Note

Cross-sectional time-series data stored as shown above are said to be in the long form. Perhaps your data are in the wide form with one observation per unit and multiple variables for the value in each year. For instance, a piece of the pulmonary function data might be

```
 pid  sex  fev91  fev92  fev93  age91  age92  age93
1071    1   1.21   1.52   1.32     25     26     28
1072    1   1.33   1.18   1.19     18     20     21
```

Data in this form can be converted to the long form by reshape; see [R] **reshape**.

❑

▷ Example

Data for some of the time periods might be missing. That is, we have cross-sectional time-series data on $i = 1, \ldots, n$ and $t = 1, \ldots, T$, but only T_i of those observations are defined. With such missing periods—called unbalanced data—a piece of our pulmonary function data might be

```
. list in 1/6, separator(0) divider
```

	pid	yr_visit	fev	age	sex	height	smokes
1.	1071	1991	1.21	25	1	69	0
2.	1071	1992	1.52	26	1	69	0
3.	1071	1993	1.32	28	1	68	0
4.	1072	1991	1.33	18	1	71	1
5.	1072	1993	1.19	21	1	71	0
6.	1073	1991	1.47	24	0	64	0

Note that patient id 1072 is not observed in 1992. The xt commands are robust to this problem.

◁

❑ Technical Note

In many of the [XT] **xt** entries, we will use data from a subsample of the NLSY data (Center for Human Resource Research 1989) on young women aged 14–26 in 1968. Women were surveyed in each of the 21 years 1968 through 1988, except for the six years 1974, 1976, 1979, 1981, 1984, and 1986. We use two different subsets: nlswork.dta and union.dta.

For nlswork.dta, our subsample is of 4,711 women in years when employed, not enrolled in school and evidently having completed their education, and with wages in excess of $1/hour but less than $700/hour.

```
. use http://www.stata-press.com/data/r8/nlswork, clear
(National Longitudinal Survey.  Young Women 14-26 years of age in 1968)

. describe
Contains data from http://www.stata-press.com/data/r8/nlswork.dta
  obs:        28,534                          National Longitudinal Survey.
                                              Young Women 14-26 years of age
                                              in 1968
  vars:          21                           9 Jun 2002 17:36
  size:   1,055,758 (89.9% of memory free)
```

	storage	display	value	
variable name	type	format	label	variable label
idcode	int	%8.0g		NLS id
year	byte	%8.0g		interview year
birth_yr	byte	%8.0g		birth year
age	byte	%8.0g		age in current year
race	byte	%8.0g		1=white, 2=black, 3=other
msp	byte	%8.0g		1 if married, spouse present
nev_mar	byte	%8.0g		1 if never yet married
grade	byte	%8.0g		current grade completed
collgrad	byte	%8.0g		1 if college graduate
not_smsa	byte	%8.0g		1 if not SMSA
c_city	byte	%8.0g		1 if central city
south	byte	%8.0g		1 if south
ind_code	byte	%8.0g		industry of employment
occ_code	byte	%8.0g		occupation
union	byte	%8.0g		1 if union
wks_ue	byte	%8.0g		weeks unemployed last year
ttl_exp	float	%9.0g		total work experience
tenure	float	%9.0g		job tenure, in years
hours	int	%8.0g		usual hours worked
wks_work	int	%8.0g		weeks worked last year
ln_wage	float	%9.0g		ln(wage/GNP deflator)

```
Sorted by:  idcode  year
```

(Continued on next page)

```
. summarize
```

Variable	Obs	Mean	Std. Dev.	Min	Max
idcode	28534	2601.284	1487.359	1	5159
year	28534	77.95865	6.383879	68	88
birth_yr	28534	48.08509	3.012837	41	54
age	28510	29.04511	6.700584	14	46
race	28534	1.303392	.4822773	1	3
msp	28518	.6029175	.4893019	0	1
nev_mar	28518	.2296795	.4206341	0	1
grade	28532	12.53259	2.323905	0	18
collgrad	28534	.1680451	.3739129	0	1
not_smsa	28526	.2824441	.4501961	0	1
c_city	28526	.357218	.4791882	0	1
south	28526	.4095562	.4917605	0	1
ind_code	28193	7.692973	2.994025	1	12
occ_code	28413	4.777672	3.065435	1	13
union	19238	.2344319	.4236542	0	1
wks_ue	22830	2.548095	7.294463	0	76
ttl_exp	28534	6.215316	4.652117	0	28.88461
tenure	28101	3.123836	3.751409	0	25.91667
hours	28467	36.55956	9.869623	1	168
wks_work	27831	53.98933	29.03232	0	104
ln_wage	28534	1.674907	.4780935	0	5.263916

For union.dta, our subset was sampled only from those with union membership information from 1970 to 1988. Our subsample is of 4,434 women. The important variables are age (16–46), grade (years of schooling completed, ranging from 0 to 18), not_smsa (28% of the person-time was spent living outside an SMSA—standard metropolitan statistical area), south (41% of the person-time was in the South), and southXt (south interacted with year, treating 1970 as year 0). You also have variable union. Overall, 22% of the person-time is marked as time under union membership, and 44% of these women have belonged to a union.

```
. use http://www.stata-press.com/data/r8/union, clear
(NLS Women 14-24 in 1968)
. describe
Contains data from http://www.stata-press.com/data/r8/union.dta
  obs:        26,200                          NLS Women 14-24 in 1968
  vars:           10                          29 Apr 2002 11:18
  size:      393,000 (96.3% of memory free)
```

variable name	storage type	display format	value label	variable label
idcode	int	%8.0g		NLS id
year	byte	%8.0g		interview year
age	byte	%8.0g		age in current year
grade	byte	%8.0g		current grade completed
not_smsa	byte	%8.0g		1 if not SMSA
south	byte	%8.0g		1 if south
union	byte	%8.0g		1 if union
t0	byte	%9.0g		
southXt	byte	%9.0g		
black	byte	%8.0g		race black

```
Sorted by:
```

```
. summarize
    Variable │      Obs        Mean    Std. Dev.        Min         Max
─────────────┼─────────────────────────────────────────────────────────
      idcode │    26200    2611.582    1484.994          1        5159
        year │    26200    79.47137    5.965499         70          88
         age │    26200    30.43221    6.489056         16          46
       grade │    26200    12.76145    2.411715          0          18
    not_smsa │    26200   .2837023    .4508027          0           1
─────────────┼─────────────────────────────────────────────────────────
       south │    26200   .4130153    .4923849          0           1
       union │    26200   .2217939    .4154611          0           1
          t0 │    26200    9.471374    5.965499          0          18
      southXt │    26200    3.96874    6.057208          0          18
       black │    26200    .274542    .4462917          0           1
```

With both datasets, we have typed

    . iis idcode
    . tis year

❑ Technical Note

The `tis` and `iis` commands, as well as other `xt` commands that set the t and i index for `xt` data, do so by declaring them as characteristics of the data; see [P] **char**. In particular, `tis` sets the characteristic `_dta[tis]` to the name of the t index variable. `iis` sets the characteristic `_dta[iis]` to the name of the i index variable.

❑ Technical Note

Throughout the `xt` entries, when random-effects models are fitted, a likelihood-ratio test that the variance of the random effects is zero is included. These tests occur on the boundary of the parameter space, invalidating the usual theory associated with such tests. However, these likelihood-ratio tests have been modified to be valid on the boundary. In particular, the null distribution of the likelihood-ratio test statistic is not the usual χ_1^2, but rather a 50:50 mixture of a χ_0^2 (point mass at zero) and a χ_1^2, denoted as $\overline{\chi}_{01}^2$. See Gutierrez et al. (2001) for a full discussion.

References

Center for Human Resource Research. 1989. *National Longitudinal Survey of Labor Market Experience, Young Women 14–26 years of age in 1968*. Ohio State University.

Gutierrez, R. G., S. L. Carter, and D. M. Drukker. 2001. sg160: On boundary-value likelihood-ratio tests. *Stata Technical Bulletin* 60: 15–18. Reprinted in *Stata Technical Bulletin Reprints*, vol. 10, pp. 269–273.

Also See

Complementary:	[XT] **xtabond**, [XT] **xtcloglog**, [XT] **xtdata**, [XT] **xtdes**, [XT] **xtfrontier**,
	[XT] **xtgee**, [XT] **xtgls**, [XT] **xthtaylor**, [XT] **xtintreg**, [XT] **xtivreg**,
	[XT] **xtlogit**, [XT] **xtnbreg**, [XT] **xtpcse**, [XT] **xtpoisson**, [XT] **xtprobit**,
	[XT] **xtrchh**, [XT] **xtreg**, [XT] **xtregar**, [XT] **xtsum**, [XT] **xttab**,
	[XT] **xttobit**,
	[TS] **tsset**

Title

> **quadchk** — Check sensitivity of quadrature approximation

Syntax

quadchk $\left[\#_1 \left[\#_2 \right] \right]$ $\left[, \underline{noout}put \right]$

Description

quadchk checks the quadrature approximation used in the random-effects estimators of the following commands:

> xtcloglog
> xtintreg
> xtlogit
> xtpoisson with the normal option
> xtprobit
> xttobit

quadchk refits the model, starting from the converged answer, for different numbers of quadrature points and then compares the different solutions.

$\#_1$ and $\#_2$ specify the number of quadrature points to use in the comparison runs of the previous model. The default is to use $n_q - 4$ and $n_q + 4$ points, where n_q is the number of quadrature points used in the original estimation.

Options

nooutput suppresses the iteration log and output of the refitted models.

Remarks

Some random-effects estimators in Stata use Gauss–Hermite quadrature to compute the log likelihood and its derivatives. The quadchk command provides a means to look at the numerical soundness of the quadrature approximation.

Using the converged coefficients of the original model as starting values, the estimation is rerun using two different numbers of quadrature points. The log likelihood and coefficient estimates for the original model and the two refitted models are then compared. If the quadrature approach is not valid, then the number of quadrature points will affect the stability of the estimation results. This instability will result in the refitted models' log likelihoods and coefficient estimates differing, sometimes dramatically, from the original model's results.

As a rule of thumb, if the coefficients do not change by more than a relative difference of 10^{-4} (0.01%), then the choice of quadrature points does not significantly affect the outcome, and the results may be confidently interpreted. However, if the results do change appreciably—greater than a relative difference of 10^{-2} (1%)—then one must question whether the model can be reliably fitted using the quadrature approach.

9

Two aspects of random-effects models have the potential to make the quadrature approximation inaccurate: large group sizes and large correlations within groups. These factors can also work in tandem, decreasing or increasing the reliability of the quadrature. For example, if the within-group correlation ρ is small, say, $\rho < 0.25$, then Gauss–Hermite quadrature may be reliable for group sizes as big as 50–100. However, when ρ is large, say greater than 0.4, simulations have shown that the quadrature can break down for group sizes as small as 20.

It is easy to see why the quadrature breaks down when group sizes are large or when ρ is big. The likelihood for a group is an integral of a normal density (the distribution of the random effects) times a product of cumulative normals. There are T_i cumulative normals in the product, where T_i is the number of observations in the ith group. The Gauss–Hermite quadrature procedure is based on the assumption that the product of normals can be approximated by a polynomial. When T_i is large or ρ is big, this assumption is no longer valid.

Note that when this assumption breaks down badly, increasing the number of quadrature points will not solve the problem. Increasing the number of quadrature points is equivalent to increasing the degree of the polynomial approximation. However, the points are positioned according to a set formula. When the number of points is increased, the range spanned by the points is also increased, and, on average, the points are only slightly closer together. If the true function is, for instance, very concentrated around zero, increasing the number of points is of little consequence because the additional points will mostly pick up the shape of the function far from zero.

When quadchk shows that the coefficient estimates change appreciably with different numbers of quadrature points, this indicates that the polynomial approximation is poor, and increasing the number of quadrature points will not help. You can convince yourself of this by continuing to increase the number of quadrature points. As you do this, the coefficient estimates will continue to change. In cases such as this, all coefficient estimates should be viewed with suspicion; one cannot claim that the results produced with larger numbers of quadrature points are more accurate than those produced with fewer points.

Simulations have shown that estimates of coefficients of independent variables that are constant within groups are especially prone to numerical instability. Hence, if your model involves independent variables of this sort, then it is especially important to run quadchk.

If the quadchk command indicates that the estimation results are sensitive to the number of quadrature points, one may want to consider an alternative such as a fixed-effects, pooled, or population-averaged model. Alternatively, if a different random-effects model is available that is not fitted via quadrature (e.g., xtpoisson, re), then that model may be a better choice.

▷ Example

In this example, we synthesize data according to the model

$$y = 0.05\,\text{x1} + 0.08\,\text{x2} + 0.08\,\text{x3} + 0.1\,\text{x4} + 0.1\,\text{x5} + 0.1\,\text{x6} + 0.1$$

$$\text{z} = \begin{cases} 1 & \text{if } y \geq 0 \\ 0 & \text{if } y < 0 \end{cases}$$

where the intrapanel correlation is 0.5 and the x1 variable is constant within panel. We first fit a random-effects probit model, and then we check the stability of the quadrature calculation:

```
. use http://www.stata-press.com/data/r8/quad1

. xtprobit z x1-x6, i(id)

Fitting comparison model:

Iteration 0:   log likelihood = -4152.5328
Iteration 1:   log likelihood = -4138.4434
Iteration 2:   log likelihood = -4138.4431

Fitting full model:

rho = 0.0      log likelihood = -4138.4431
rho = 0.1      log likelihood =   -3603.06
rho = 0.2      log likelihood = -3448.0667
rho = 0.3      log likelihood =  -3382.909
rho = 0.4      log likelihood = -3356.2536
rho = 0.5      log likelihood = -3354.0627
rho = 0.6      log likelihood = -3376.4348
Iteration 0:   log likelihood = -3354.0627
Iteration 1:   log likelihood = -3352.1745
Iteration 2:   log likelihood = -3349.6987
Iteration 3:   log likelihood = -3349.6926
```

Random-effects probit regression				Number of obs	=	6000
Group variable (i): id				Number of groups	=	300
Random effects u_i ~ Gaussian				Obs per group: min =		20
				avg =		20.0
				max =		20
				Wald chi2(6)	=	36.15
Log likelihood = -3349.6926				Prob > chi2	=	0.0000

z	Coef.	Std. Err.	z	P>|z|	[95% Conf. Interval]	
x1	.1156763	.0554911	2.08	0.037	.0069157	.2244369
x2	.1005555	.066227	1.52	0.129	-.0292469	.230358
x3	.1542187	.0660852	2.33	0.020	.0246942	.2837432
x4	.1257616	.0375776	3.35	0.001	.0521109	.1994123
x5	.1366003	.0654695	2.09	0.037	.0082824	.2649182
x6	.0870325	.0453489	1.92	0.055	-.0018496	.1759147
_cons	.1098393	.0500502	2.19	0.028	.0117426	.2079359
/lnsig2u	-.0791821	.0971059			-.2695062	.1111419
sigma_u	.9611824	.0466682			.8739317	1.057144
rho	.4802148	.0242385			.4330283	.5277569

```
Likelihood-ratio test of rho=0: chibar2(01) =  1577.50 Prob >= chibar2 = 0.000

. quadchk

Refitting model quad() =  8
  (output omitted)

Refitting model quad() = 16
  (output omitted)
```

(Continued on next page)

Quadrature check

	Fitted quadrature 12 points	Comparison quadrature 8 points	Comparison quadrature 16 points	
Log likelihood	-3349.6926	-3354.6372 -4.9446636 .00147615	-3348.3881 1.3045064 -.00038944	 Difference Relative difference
z: x1	.11567632	.16153997 .04586365 .3964826	.07007833 -.04559799 -.39418607	 Difference Relative difference
z: x2	.10055552	.10317831 .00262279 .02608296	.09937417 -.00118135 -.01174825	 Difference Relative difference
z: x3	.1542187	.15465369 .00043499 .00282062	.15150516 -.00271354 -.0175954	 Difference Relative difference
z: x4	.12576159	.12880254 .00304096 .02418032	.1243974 -.00136418 -.01084739	 Difference Relative difference
z: x5	.13660028	.13475211 -.00184817 -.01352977	.13707075 .00047047 .00344411	 Difference Relative difference
z: x6	.08703252	.08568342 -.0013491 -.0155011	.08738135 .00034883 .00400808	 Difference Relative difference
z: _cons	.10983928	.11031299 .00047371 .00431278	.09654975 -.01328953 -.12099065	 Difference Relative difference
lnsig2u: _cons	-.07918213	-.18133823 -.1021561 1.2901408	-.05815644 .02102569 -.26553574	 Difference Relative difference

We see that the x1 variable (the one that was constant within panel) changed with a relative difference of nearly 40%! Hence, we conclude that we cannot trust the quadrature approximation for this model, and all results are considered suspect.

◁

▷ Example

In this example, we synthesize data exactly the same way as in the previous example, but we make the intrapanel correlation equal to 0.1 instead of 0.5. We again fit a random-effects probit model and check the quadrature:

(Continued on next page)

```
. use http://www.stata-press.com/data/r8/quad2

. xtprobit z x1-x6, i(id) nolog
```

```
Random-effects probit regression              Number of obs      =      6000
Group variable (i): id                        Number of groups   =       300

Random effects u_i ~ Gaussian                 Obs per group: min =        20
                                                             avg =      20.0
                                                             max =        20

                                              Wald chi2(6)       =     39.43
Log likelihood  = -4065.3144                  Prob > chi2        =    0.0000
```

z	Coef.	Std. Err.	z	P>|z|	[95% Conf. Interval]	
x1	.0246934	.0251121	0.98	0.325	-.0245255	.0739123
x2	.1300122	.0587907	2.21	0.027	.0147847	.2452398
x3	.1190411	.0579539	2.05	0.040	.0054535	.2326287
x4	.1391966	.0331817	4.19	0.000	.0741617	.2042316
x5	.0773645	.0578455	1.34	0.181	-.0360106	.1907395
x6	.0862025	.0401185	2.15	0.032	.0075716	.1648334
_cons	.0922659	.0244394	3.78	0.000	.0443656	.1401661
/lnsig2u	-2.34394	.1575243			-2.652682	-2.035198
sigma_u	.3097561	.0243971			.2654468	.3614618
rho	.0875487	.0125837			.0658239	.1155566

```
Likelihood-ratio test of rho=0: chibar2(01) =    110.19 Prob >= chibar2 = 0.000

. quadchk, nooutput

Refitting model quad() =  8
Refitting model quad() = 16
```

```
                              Quadrature check

                    Fitted        Comparison     Comparison
                    quadrature    quadrature     quadrature
                    12 points     8 points       16 points
```

	Fitted quadrature 12 points	Comparison quadrature 8 points	Comparison quadrature 16 points	
Log likelihood	-4065.3144	-4065.3173	-4065.3144	
		-.00286401	-4.767e-06	Difference
		7.045e-07	1.172e-09	Relative difference
z: x1	.02469338	.02468991	.02469426	
		-3.463e-06	8.851e-07	Difference
		-.00014023	.00003584	Relative difference
z: x2	.13001225	.13001198	.13001229	
		-2.663e-07	4.027e-08	Difference
		-2.048e-06	3.097e-07	Relative difference
z: x3	.11904112	.11901865	.1190409	
		-.00002247	-2.199e-07	Difference
		-.00018879	-1.847e-06	Relative difference
z: x4	.13919664	.13918545	.13919696	
		-.00001119	3.232e-07	Difference
		-.00008037	2.322e-06	Relative difference
z: x5	.07736447	.0773757	.07736399	
		.00001123	-4.849e-07	Difference
		.00014516	-6.268e-06	Relative difference

z:				
x6	.0862025	.08618573	.08620282	
		-.00001677	3.264e-07	Difference
		-.00019454	3.786e-06	Relative difference
z:				
_cons	.09226589	.09224255	.09226531	
		-.00002334	-5.753e-07	Difference
		-.00025297	-6.236e-06	Relative difference
lnsig2u:				
_cons	-2.3439398	-2.3442475	-2.3439384	
		-.00030763	1.450e-06	Difference
		.00013124	-6.187e-07	Relative difference

Here we see that the quadrature approximation is stable, even for the coefficient of x1. With this result, you can confidently interpret the results.

Again, note that the only difference between this example and the previous one is the value of ρ. The quadrature approximation works wonderfully for small to moderate values of ρ, but it breaks down for large values of ρ. Indeed, for large values of ρ, one should do more than question the validity of the quadrature approximation; one should question the validity of the random-effects model itself.

◁

Methods and Formulas

quadchk is implemented as an ado-file.

Also See

Complementary: [XT] **xtcloglog**, [XT] **xtintreg**, [XT] **xtlogit**, [XT] **xtpoisson**,
[XT] **xtprobit**, [XT] **xttobit**

Title

> **xtabond** — Arellano–Bond linear, dynamic panel-data estimation

Syntax

> xtabond *depvar* [*varlist*] [if *exp*] [in *range*] [, <u>lags</u>(#) <u>maxldep</u>(#)
>
>   <u>maxlags</u>(#) <u>diff</u>vars(*varlist*) inst(*varlist*)
>
>   pre(*varlist* [, lagstruct(#,#) <u>end</u>ogenous])
>
>   [pre(*varlist* [, lagstruct(#,#) <u>end</u>ogenous])
>
>   ... [pre(*varlist* [, lagstruct(#,#) <u>end</u>ogenous])]
>
>   <u>ar</u>tests(#) <u>r</u>obust <u>tw</u>ostep <u>no</u>constant <u>sm</u>all <u>l</u>evel(#)]

by ... : may be used with xtabond; see [R] **by**.

xtabond is for use with time-series data; see [TS] **tsset**. You must tsset your data before using xtabond.

All *varlist*s may contain time-series operators; see [U] **14.4.3 Time-series varlists**. The specification of *depvar*, however, may not contain time-series operators.

xtabond shares the features of all estimation commands; see [U] **23 Estimation and post-estimation commands**.

Syntax for predict

> predict [*type*] *newvarname* [if *exp*] [in *range*] [, *statistic*]

where *statistic* is

xb	$a + \mathbf{x}_{it}\mathbf{b}$, fitted values (the default)	
e	e_{it}, the residuals	

Description

xtabond fits a dynamic panel-data model via the Arellano–Bond estimator. Consider the model

$$y_{it} = \sum_{j=1}^{p} \alpha_j y_{i,t-j} + \mathbf{x}_{it}\boldsymbol{\beta}_1 + \mathbf{w}_{it}\boldsymbol{\beta}_2 + \nu_i + \epsilon_{it} \quad i = 1, \ldots, N \quad t = 1, \ldots, T_i \quad (1)$$

where

the α_j are p parameters to be estimated

$\mathbf{x}_{it}$ is a $1 \times k_1$ vector of strictly exogenous covariates

$\boldsymbol{\beta}_1$ is a $k_1 \times 1$ vector of parameters to be estimated

$\mathbf{w}_{it}$ is a $1 \times k_2$ vector of predetermined covariates

$\boldsymbol{\beta}_2$ is a $k_2 \times 1$ vector of parameters to be estimated

ν_i are the random effects that are independent and identically distributed (i.i.d.) over the panels with variance σ_ν^2

and ϵ_{it} are i.i.d. over the whole sample with variance σ_ϵ^2.

It is also assumed that the ν_i and the ϵ_{it} are independent for each i over all t.

First differencing equation (1) removes the ν_i and produces an equation that is estimable by instrumental variables. Arellano and Bond (1991) derived a generalized method of moments estimator for α_j, $j \in \{1, \ldots, p\}$, $\boldsymbol{\beta}_1$, and $\boldsymbol{\beta}_2$ using lagged levels of the dependent variable and the predetermined variables and differences of the strictly exogenous variables. xtabond implements this estimator, known as the Arellano–Bond dynamic panel-data estimator. This methodology assumes that there is no second-order autocorrelation in the first-differenced idiosyncratic errors. xtabond includes the test for autocorrelation and the Sargan test of over-identifying restrictions for this model derived by Arellano and Bond (1991).

Options

lags(#) sets p, the number of lags of the dependent variable to be included in the model. The default is $p = 1$.

maxldep(#) sets the maximum number of lags of the dependent variable that can be used as instruments. The default is to use all $T_i - p - 2$ lags.

maxlags(#) sets the maximum number of lags of the dependent variable and of all the predetermined variables that can be used as instruments. The default is to use all $T_i - p - 2$ lags of the dependent variable. If the predetermined variables are endogenous, the default is to use all $T_i - p - 2$ lags of these endogenous variables. If the predetermined variables are not endogenous, the default is to use all $T_i - p - 1$ lags of these variables.

diffvars(*varlist*) specifies a set of variables that have already been differenced to be included as strictly exogenous covariates.

inst(*varlist*) specifies a set of variables to be used as additional instruments. These instruments are not differenced by xtabond before including them into the instrument matrix.

pre(*varlist* [, lagstruct(*prelags*, *premaxlags*) endogenous]) specifies that a set of predetermined variables is to be included in the model. Optionally, one may specify that *prelags* lags of the specified variables are also included. The default for *prelags* is 0. Specifying *premaxlags* sets the maximum number of further lags of the predetermined variables to be used as instruments. Additionally, specifying the endogenous option causes xtabond to treat these variables as endogenous instead of just predetermined. The default is to include $T_i - prelags - 1$ lagged levels as instruments for predetermined variables and $T_i - prelags - 2$ lagged levels as instruments for endogenous variables. The user may specify as many sets of predetermined variables as is feasible within the standard Stata limits on matrix size. Each set of predetermined variables may have its own number of *prelags* and *premaxlags*.

artests(#) specifies the maximum order of the autocorrelation test to be calculated and reported. The maximum order must be less than or equal to $p + 1$. The default is 2.

robust specifies that the robust estimator of the variance–covariance matrix of the parameter estimates is to be calculated and reported. This option is not available with two-step estimates.

twostep specifies that the two-step estimator is to be calculated and reported.

noconstant suppresses the constant term (intercept) in the regression.

small specifies that t statistics should be reported instead of Z statistics, and that F statistics should be reported instead of chi-squared statistics.

level(#) specifies the confidence level, in percent, for confidence intervals. The default is level(95) or as set by set level; see [U] **23.6 Specifying the width of confidence intervals**.

Options for predict

xb, the default, calculates the linear prediction from the first-differenced equation.

e calculates the residual error of the differenced dependent variable from the linear prediction.

Remarks

Anderson and Hsiao (1981, 1982) proposed using further lags of the level or the difference of the dependent variable to instrument the lagged dependent variables included in a dynamic panel-data model after the random effects had been removed by first differencing. A version of this estimator can be obtained from xtivreg. See [XT] **xtivreg** for an example. Arellano and Bond (1991) built on this idea by noting that, in general, there are many more instruments available. Using the GMM framework developed by Hansen (1982), they identified how many lags of the dependent variable and the predetermined variables were valid instruments, and how to combine these lagged levels with first differences of the strictly exogenous variables into a potentially very large instrument matrix. Using this instrument matrix, Arellano and Bond (1991) derived the corresponding one-step and two-step GMM estimators. They also found the robust VCE estimator for the one-step model. In addition, they derived a test of autocorrelation of order m and the Sargan test of over-identifying restrictions for this estimator.

▷ Example

In their article, Arellano and Bond (1991) applied their new estimators and test statistics to a model of dynamic labor demand that had previously been considered by Layard and Nickell (1986). They use data from an unbalanced panel of firms from the United Kingdom. As is conventional, all variables are indexed over the firm i and time t. In this dataset, n_{it} is the log of employment in firm i inside the U.K. at time t, w_{it} is the natural log of the real product wage, k_{it} is the natural log of the gross capital stock, and ys_{it} is the natural log of industry output. The model also includes time dummies yr1980, yr1981, yr1982, yr1983, and yr1984. In Table 4 of Arellano and Bond (1991), the authors present the results they obtained from several specifications.

In column (a1) of Table 4, Arellano and Bond report the coefficients and their standard errors from the robust one-step estimators of a dynamic model of labor demand. In order to clarify some important issues, we will begin with the homoskedastic one-step version of this model, and then consider the robust case. Here is the command using xtabond and the subsequent output for the homoskedastic case:

```
. use http://www.stata-press.com/data/r8/abdata

. xtabond n l(0/1).w l(0/2).(k ys) yr1980-yr1984, lags(2)
```

```
Arellano-Bond dynamic panel-data estimation       Number of obs      =       611
Group variable (i): id                            Number of groups   =       140

                                                  Wald chi2(15)      =    575.84

Time variable (t): year                           Obs per group: min =         4
                                                                 avg =  4.364286
                                                                 max =         6
```

One-step results

n		Coef.	Std. Err.	z	P>|z|	[95% Conf. Interval]	
n							
	LD	.6862262	.1486163	4.62	0.000	.3949435	.9775088
	L2D	-.0853582	.0444365	-1.92	0.055	-.1724523	.0017358
w							
	D1	-.6078208	.0657694	-9.24	0.000	-.7367265	-.4789151
	LD	.3926237	.1092374	3.59	0.000	.1785222	.6067251
k							
	D1	.3568456	.0370314	9.64	0.000	.2842653	.4294259
	LD	-.0580012	.0583051	-0.99	0.320	-.172277	.0562747
	L2D	-.0199475	.0416274	-0.48	0.632	-.1015357	.0616408
ys							
	D1	.6085073	.1345412	4.52	0.000	.3448115	.8722031
	LD	-.7111651	.1844599	-3.86	0.000	-1.0727	-.3496304
	L2D	.1057969	.1428568	0.74	0.459	-.1741974	.3857912
yr1980							
	D1	.0029062	.0212705	0.14	0.891	-.0387832	.0445957
yr1981							
	D1	-.0404378	.0354707	-1.14	0.254	-.1099591	.0290836
yr1982							
	D1	-.0652767	.048209	-1.35	0.176	-.1597646	.0292111
yr1983							
	D1	-.0690928	.0627354	-1.10	0.271	-.1920521	.0538664
yr1984							
	D1	-.0650302	.0781322	-0.83	0.405	-.2181665	.0881061
_cons		.0095545	.0142073	0.67	0.501	-.0182912	.0374002

```
Sargan test of over-identifying restrictions:
       chi2(25) =      65.82     Prob > chi2 = 0.0000

Arellano-Bond test that average autocovariance in residuals of order 1 is 0:
       H0: no autocorrelation   z =  -3.94   Pr > z = 0.0001
Arellano-Bond test that average autocovariance in residuals of order 2 is 0:
       H0: no autocorrelation   z =  -0.54   Pr > z = 0.5876
```

The coefficients are identical to those reported in column (a1) of Table 4, as they should be. Of course, the standard errors are different because we are considering the homoskedastic case. Only in the case of a homoskedastic error term does the Sargan test have an asymptotic chi-squared distribution. In fact, Arellano and Bond (1991) found evidence that the one-step Sargan test over-rejects in the presence of heteroskedasticity. Since its asymptotic distribution is not known under the assumptions of the robust model, xtabond will not compute it when robust is specified. The Sargan test, reported by Arellano and Bond (1991, Table 4, column a1), comes from the one-step homoskedastic estimator, and is the same as the one reported here. By default, xtabond calculates and reports the Arellano–Bond test for first and second-order autocorrelation in the first-differenced residuals. There are versions of this test for both the homoskedastic and the robust cases, although their values are different.

◁

▷ Example

Now consider the output from the one-step robust estimator of the same model:

```
. xtabond n l(0/1).w l(0/2).(k ys) yr1980-yr1984, lags(2) robust
```

```
Arellano-Bond dynamic panel-data estimation     Number of obs      =        611
Group variable (i): id                          Number of groups   =        140

                                                Wald chi2(15)      =     618.58

Time variable (t): year                         Obs per group: min =          4
                                                               avg =   4.364286
                                                               max =          6
```

One-step results

n		Coef.	Robust Std. Err.	z	P>|z|	[95% Conf. Interval]	
n							
	LD	.6862262	.1445943	4.75	0.000	.4028266	.9696257
	L2D	-.0853582	.0560155	-1.52	0.128	-.1951467	.0244302
w							
	D1	-.6078208	.1782055	-3.41	0.001	-.9570972	-.2585445
	LD	.3926237	.1679931	2.34	0.019	.0633632	.7218842
k							
	D1	.3568456	.0590203	6.05	0.000	.241168	.4725233
	LD	-.0580012	.0731797	-0.79	0.428	-.2014308	.0854284
	L2D	-.0199475	.0327126	-0.61	0.542	-.0840631	.0441681
ys							
	D1	.6085073	.1725313	3.53	0.000	.2703522	.9466624
	LD	-.7111651	.2317163	-3.07	0.002	-1.165321	-.2570095
	L2D	.1057969	.1412021	0.75	0.454	-.1709542	.382548
yr1980							
	D1	.0029062	.0158028	0.18	0.854	-.0280667	.0338791
yr1981							
	D1	-.0404378	.0280582	-1.44	0.150	-.0954307	.0145552
yr1982							
	D1	-.0652767	.0365451	-1.79	0.074	-.1369038	.0063503
yr1983							
	D1	-.0690928	.047413	-1.46	0.145	-.1620205	.0238348
yr1984							
	D1	-.0650302	.0576305	-1.13	0.259	-.1779839	.0479235
_cons		.0095545	.0102896	0.93	0.353	-.0106127	.0297217

```
Arellano-Bond test that average autocovariance in residuals of order 1 is 0:
        H0: no autocorrelation   z =  -3.60   Pr > z = 0.0003
Arellano-Bond test that average autocovariance in residuals of order 2 is 0:
        H0: no autocorrelation   z =  -0.52   Pr > z = 0.6058
```

The coefficients are the same, but not the standard errors, and the value of the test for second-order autocorrelation matches that reported in Arellano and Bond (1991, Table 4, column a1). As one might suspect, most of the robust standard errors are higher than those that assume a homoskedastic error term. Note that xtabond does not report the Sargan statistic in this case.

Overall, the test results from the one-step model are mixed. The Sargan test from the one-step homoskedastic estimator rejects the null hypothesis that the over-identifying restrictions are valid, but this could be due to heteroskedasticity. In both the homoskedastic and the robust cases, the null hypothesis of no first-order autocorrelation in the differenced residuals is rejected, but it is not possible to reject the null hypothesis of no second-order autocorrelation. The presence of first-order autocorrelation in the differenced residuals does not imply that the estimates are inconsistent, but the presence of second-order autocorrelation would imply that the estimates are inconsistent. (See

Arellano and Bond (1991, 281–282) for a discussion on this point.) The above output indicates that we have included several statistically insignificant variables. However, since we want the results to change with the estimation procedure for a given model, we will not remove these variables.

◁

▷ Example

xtabond reports the Wald statistic of the null hypothesis that all the coefficients except the constant are zero. For the case at hand, the null hypothesis is soundly rejected. In column (a1) of Table 4, Arellano and Bond report a chi-squared test of the null hypothesis that all the coefficients, except the constant and the time dummies, are zero. Here is an example of how to perform this test in Stata:

```
. test ld.n l2d.n d.w ld.w d.k ld.k l2d.k d.ys ld.ys l2d.ys

 ( 1)  LD.n = 0
 ( 2)  L2D.n = 0
 ( 3)  D.w = 0
 ( 4)  LD.w = 0
 ( 5)  D.k = 0
 ( 6)  LD.k = 0
 ( 7)  L2D.k = 0
 ( 8)  D.ys = 0
 ( 9)  LD.ys = 0
 (10)  L2D.ys = 0

          chi2( 10) =   408.29
        Prob > chi2 =    0.0000
```

Since the rejection of the null hypothesis of the Sargan test may indicate the presence of heteroskedasticity, we might expect large efficiency gains from using the two-step estimator. The two-step estimator of the same model produces the following:

(Continued on next page)

```
. xtabond n l(0/1).w l(0/2).(k ys) yr1980-yr1984, lags(2) twostep
```

```
Arellano-Bond dynamic panel-data estimation      Number of obs       =        611
Group variable (i): id                           Number of groups    =        140

                                                 Wald chi2(15)       =    1035.56

Time variable (t): year                          Obs per group: min  =          4
                                                                avg  =   4.364286
                                                                max  =          6
```

```
Two-step results
```

n		Coef.	Std. Err.	z	P>|z|	[95% Conf. Interval]	
n							
	LD	.6287089	.0904543	6.95	0.000	.4514216	.8059961
	L2D	-.0651882	.0265009	-2.46	0.014	-.117129	-.0132474
w							
	D1	-.5257597	.0537692	-9.78	0.000	-.6311453	-.420374
	LD	.3112899	.0940116	3.31	0.001	.1270305	.4955492
k							
	D1	.2783619	.0449083	6.20	0.000	.1903432	.3663807
	LD	.0140994	.0528046	0.27	0.789	-.0893957	.1175946
	L2D	-.0402484	.0258038	-1.56	0.119	-.0908229	.010326
ys							
	D1	.5919243	.1162114	5.09	0.000	.3641542	.8196943
	LD	-.5659863	.1396738	-4.05	0.000	-.8397419	-.2922306
	L2D	.1005433	.1126749	0.89	0.372	-.1202955	.321382
yr1980							
	D1	.0006378	.0127959	0.05	0.960	-.0244417	.0257172
yr1981							
	D1	-.0550044	.0235162	-2.34	0.019	-.1010953	-.0089135
yr1982							
	D1	-.075978	.0302659	-2.51	0.012	-.135298	-.0166579
yr1983							
	D1	-.0740708	.0370993	-2.00	0.046	-.146784	-.0013575
yr1984							
	D1	-.0906606	.0453924	-2.00	0.046	-.179628	-.0016933
_cons		.0112155	.0077507	1.45	0.148	-.0039756	.0264066

```
Warning: Arellano and Bond recommend using one-step results for
         inference on coefficients

Sargan test of over-identifying restrictions:
         chi2(25) =     31.38     Prob > chi2 = 0.1767

Arellano-Bond test that average autocovariance in residuals of order 1 is 0:
         H0: no autocorrelation   z =  -3.00   Pr > z = 0.0027
Arellano-Bond test that average autocovariance in residuals of order 2 is 0:
         H0: no autocorrelation   z =  -0.42   Pr > z = 0.6776
```

Note that Arellano and Bond recommend using the one-step results for inference on the coefficients. Several studies have found that the two-step standard errors tend to be biased downward in small samples. For this reason, the one-step results are generally recommended for inference. (See Arellano and Bond (1991) for details.) However, as this example illustrates, the two-step Sargan test may be better for inference on model specification.

In terms of interpreting the output, the most important change is that we can no longer reject the null hypothesis in the Sargan test. However, it remains safe to reject the null hypothesis of no first-order serial correlation in the differenced residuals. Note that the magnitudes of several of the coefficient estimates have changed, and that one even switched its sign.

◁

▷ Example

In some cases, the assumption of strict exogeneity is not tenable. Recall that a variable x_{it} is said to be strictly exogenous if $E[x_{it}\epsilon_{is}] = 0$ for all t and s. If $E[x_{it}\epsilon_{is}] \neq 0$ for $s < t$ but $E[x_{it}\epsilon_{is}] = 0$ for all $s \geq t$, the variable is said to be predetermined. Intuitively, if the error term at time t has some feedback on the subsequent realizations of x_{it}, then x_{it} is a predetermined variable. Since unforecastable errors today might affect future changes in the real wage and in the capital stock, one might suspect that the log of the real product wage and the log of the gross capital stock are not strictly exogenous, but are predetermined. In this example, we treat w and k as predetermined, and use levels lagged one or more periods as instruments.

```
. xtabond n l(0/1).ys yr1980-yr1984, lags(2) twostep pre(w, lag(1,.)) pre(k,lag(2,.))
Arellano-Bond dynamic panel-data estimation      Number of obs      =       611
Group variable (i): id                           Number of groups   =       140

                                                 Wald chi2(14)      =  17241.41

Time variable (t): year                          Obs per group: min =         4
                                                                avg =  4.364286
                                                                max =         6

Two-step results
```

n		Coef.	Std. Err.	z	P>|z|	[95% Conf. Interval]	
n							
	LD	.6343155	.0121403	52.25	0.000	.610521	.6581099
	L2D	-.0871247	.0075422	-11.55	0.000	-.1019071	-.0723422
w							
	D1	-.720063	.0144539	-49.82	0.000	-.7483921	-.6917339
	LD	.238069	.028488	8.36	0.000	.1822335	.2939044
k							
	D1	.3931997	.0255093	15.41	0.000	.3432025	.443197
	LD	-.0019641	.0155638	-0.13	0.900	-.0324685	.0285403
	L2D	-.0231165	.0098741	-2.34	0.019	-.0424693	-.0037637
ys							
	D1	.5999718	.0475755	12.61	0.000	.5067257	.693218
	LD	-.5674808	.0535028	-10.61	0.000	-.6723443	-.4626173
yr1980							
	D1	-.006209	.006551	-0.95	0.343	-.0190487	.0066307
yr1981							
	D1	-.0398491	.0132912	-3.00	0.003	-.0658993	-.0137988
yr1982							
	D1	-.0525715	.0190716	-2.76	0.006	-.0899512	-.0151918
yr1983							
	D1	-.0451175	.0246485	-1.83	0.067	-.0934277	.0031927
yr1984							
	D1	-.0437772	.0285991	-1.53	0.126	-.0998304	.0122761
_cons		.0173374	.0051688	3.35	0.001	.0072066	.0274681

```
Warning: Arellano and Bond recommend using one-step results for
         inference on coefficients

Sargan test of over-identifying restrictions:
        chi2(86) =     89.45      Prob > chi2 = 0.3783
Arellano-Bond test that average autocovariance in residuals of order 1 is 0:
        H0: no autocorrelation   z =  -4.04   Pr > z = 0.0001
Arellano-Bond test that average autocovariance in residuals of order 2 is 0:
        H0: no autocorrelation   z =  -0.37   Pr > z = 0.7123
```

The increase in the p-value of the Sargan test indicates that treating the w and k as predetermined makes it more difficult to reject the null hypothesis that the over-identifying restrictions are valid.

This increase in the Sargan statistic provides some evidence that w and k are better modeled as predetermined variables.

◁

▷ Example

Alternatively, we might suspect that w and k are endogenous. Here, we mean endogenous in that $E[x_{it}\epsilon_{is}] \neq 0$ for $s \leq t$ but $E[x_{it}\epsilon_{is}] = 0$ for all $s > t$. Note that by this definition, endogenous differs from predetermined only in that the former allows for correlation between the x_{it} and the v_{it} at time t, while the latter does not. Endogenous variables are treated similarly to the lagged dependent variable. Levels of the endogenous variables lagged two or more periods are available to serve as instruments. In this example, we treat w and k as endogenous variables.

```
. xtabond n l(0/1).ys yr1980-yr1984, lags(2) twostep pre(w, lag(1,.) endog)
> pre(k,lag(2,.) endog)
```

```
Arellano-Bond dynamic panel-data estimation      Number of obs     =       611
Group variable (i): id                           Number of groups  =       140

                                                 Wald chi2(14)     =   8044.21
Time variable (t): year                          Obs per group: min =        4
                                                                avg = 4.364286
                                                                max =        6

Two-step results
```

n		Coef.	Std. Err.	z	P>|z|	[95% Conf. Interval]	
n							
	LD	.848743	.0292901	28.98	0.000	.7913355	.9061505
	L2D	-.0988995	.0163296	-6.06	0.000	-.1309049	-.0668941
w							
	D1	-.6727914	.0215607	-31.20	0.000	-.7150495	-.6305333
	LD	.5903514	.0365629	16.15	0.000	.5186894	.6620135
k							
	D1	.3886398	.0319844	12.15	0.000	.3259515	.4513282
	LD	-.1735626	.0277832	-6.25	0.000	-.2280166	-.1191086
	L2D	-.0500805	.0155138	-3.23	0.001	-.080487	-.019674
ys							
	D1	.6665421	.0608645	10.95	0.000	.54725	.7858343
	LD	-.786225	.0704774	-11.16	0.000	-.9243582	-.6480919
yr1980							
	D1	-.004441	.0082162	-0.54	0.589	-.0205445	.0116625
yr1981							
	D1	-.0533043	.0153854	-3.46	0.001	-.0834591	-.0231496
yr1982							
	D1	-.0941273	.0216129	-4.36	0.000	-.1364877	-.0517669
yr1983							
	D1	-.1154812	.0283393	-4.07	0.000	-.1710252	-.0599372
yr1984							
	D1	-.1374538	.0310743	-4.42	0.000	-.1983584	-.0765493
_cons		.0209066	.0052402	3.99	0.000	.0106359	.0311772

```
Warning: Arellano and Bond recommend using one-step results for
         inference on coefficients

Sargan test of over-identifying restrictions:
       chi2(74) =    74.92      Prob > chi2 = 0.4483

Arellano-Bond test that average autocovariance in residuals of order 1 is 0:
        H0: no autocorrelation   z =  -4.91   Pr > z = 0.0000
Arellano-Bond test that average autocovariance in residuals of order 2 is 0:
        H0: no autocorrelation   z =  -0.80   Pr > z = 0.4251
```

While some of the estimated coefficients changed in magnitude, none of the estimates changed in sign, and these results are generally very similar to those obtained by treating w and k as predetermined.

◁

▷ Example

Treating variables as predetermined, or endogenous, increases the size of the instrument matrix very quickly. (See the *Methods and Formulas* section for a discussion of how this matrix is created and what determines its size.) There are two potential problems with a very large instrument matrix. First, GMM estimators with too many over-identifying restrictions may perform poorly in small samples. (See Kiviet (1995) for a discussion of the dynamic panel-data case.) Second, the problem may become too large to estimate. The instrument matrix cannot exceed the current limit on matsize. For instance, to fit the above model, matsize must be at least 101. Here is what would have happened if we attempted the estimation with a smaller matsize:

```
. set matsize 50
. xtabond n l(0/1).ys yr1980-yr1984, lags(2) twostep pre(w, lag(1,.)) pre(k,lag(2,.))
matsize too small
    You have attempted to create a matrix with more than 50 rows or columns or
    to fit a model with more than 50 variables plus ancillary parameters.
    You need to increase matsize using the set matsize command; see help
    matsize.
matsize must be at least 101
(you have 101 instruments)
```

To handle these problems, it is conventional to allow users to set a maximum number of lagged levels to be included as instruments for the predetermined variables. Here is an example in which a maximum of three lagged levels of the predetermined variables are included as instruments:

(Continued on next page)

```
. xtabond n l(0/1).ys yr1980-yr1984, lags(2) twostep pre(w,lag(1,3)) pre(k,lag(2,3))
```

Arellano-Bond dynamic panel-data estimation   Number of obs   =   611
Group variable (i): id   Number of groups =   140

                                                Wald chi2(14)   =   4139.65
Time variable (t): year   Obs per group: min =   4
                                                             avg = 4.364286
                                                             max =   6

Two-step results

n		Coef.	Std. Err.	z	P>|z|	[95% Conf. Interval]	
n							
	LD	.6386374	.0294894	21.66	0.000	.5808392	.6964356
	L2D	-.0556839	.0167345	-3.33	0.001	-.0884829	-.022885
w							
	D1	-.7560323	.0261701	-28.89	0.000	-.8073248	-.7047398
	LD	.2596916	.0379468	6.84	0.000	.1853172	.3340659
k							
	D1	.3562148	.0419901	8.48	0.000	.2739158	.4385138
	LD	-.0114216	.020624	-0.55	0.580	-.0518439	.0290007
	L2D	-.0400761	.014737	-2.72	0.007	-.06896	-.0111922
ys							
	D1	.7043524	.0725895	9.70	0.000	.5620796	.8466251
	LD	-.5920347	.0710921	-8.33	0.000	-.7313726	-.4526968
yr1980							
	D1	.0027573	.0086318	0.32	0.749	-.0141607	.0196754
yr1981							
	D1	-.027851	.015943	-1.75	0.081	-.0590987	.0033967
yr1982							
	D1	-.0411806	.0217577	-1.89	0.058	-.0838249	.0014637
yr1983							
	D1	-.0395438	.0280304	-1.41	0.158	-.0944824	.0153948
yr1984							
	D1	-.0469629	.0319918	-1.47	0.142	-.1096657	.0157398
_cons		.0169447	.0058257	2.91	0.004	.0055265	.0283628

Warning: Arellano and Bond recommend using one-step results for
         inference on coefficients

Sargan test of over-identifying restrictions:
        chi2(66) =   72.07   Prob > chi2 = 0.2842

Arellano-Bond test that average autocovariance in residuals of order 1 is 0:
        H0: no autocorrelation   z = -4.14   Pr > z = 0.0000
Arellano-Bond test that average autocovariance in residuals of order 2 is 0:
        H0: no autocorrelation   z = -0.86   Pr > z = 0.3878

◁

(Continued on next page)

▷ Example

xtabond handles data in which there are missing observations in the middle of the panels. In the following example, we deliberately set the dependent variable to missing in the year 1980:

```
. replace n=. if year==1980
(140 real changes made, 140 to missing)
. xtabond n l(0/1).w l(0/2).(k ys) yr1980-yr1984, lags(2) robust
note: yr1981 dropped due to collinearity
note: yr1982 dropped due to collinearity
note: the residuals and the L(1) residuals have no obs in common
      The AR(1) is trivially zero
note: the residuals and the L(2) residuals have no obs in common
      The AR(2) is trivially zero
```

Arellano-Bond dynamic panel-data estimation	Number of obs = 115
Group variable (i): id	Number of groups = 101
	Wald chi2(11) = 42.19
Time variable (t): year	Obs per group: min = 1
	avg = 1.138614
	max = 2

One-step results

n		Coef.	Robust Std. Err.	z	P>|z|	[95% Conf. Interval]
n							
	LD	.1782912	.2213558	0.81	0.421	-.2555582	.6121406
	L2D	.0305248	.0524455	0.58	0.561	-.0722665	.133316
w							
	D1	-.265179	.1497362	-1.77	0.077	-.5586565	.0282986
	LD	.1861825	.1430316	1.30	0.193	-.0941542	.4665193
k							
	D1	.3971506	.0887184	4.48	0.000	.2232657	.5710354
	LD	-.0310769	.0908215	-0.34	0.732	-.2090837	.14693
	L2D	-.0374041	.0642649	-0.58	0.561	-.163361	.0885528
ys							
	D1	.3872922	.3840567	1.01	0.313	-.365445	1.14003
	LD	-.6809435	.4930012	-1.38	0.167	-1.647208	.2853211
	L2D	.5562209	.4333161	1.28	0.199	-.2930632	1.405505
yr1980							
	D1	(dropped)					
yr1983							
	D1	-.006329	.0257789	-0.25	0.806	-.0568547	.0441968
yr1984							
	D1	(dropped)					
_cons		.0015564	.0104385	0.15	0.881	-.0189027	.0220154

```
Arellano-Bond test that average autocovariance in residuals of order 1 is 0:
       H0: no autocorrelation   z =    .   Pr > z =    .
Arellano-Bond test that average autocovariance in residuals of order 2 is 0:
       H0: no autocorrelation   z =    .   Pr > z =    .
```

There are two important aspects to this example. First, note the warnings and the missing values for the AR tests. We asked xtabond to compute tests for which it did not have sufficient data, so it issued warnings and set the values of the tests to missing. Second, since xtabond uses time-series operators in its computations, if statements and missing values are not equivalent. An if statement will cause the false observations to be excluded from the sample, but it will compute the time-series operators wherever possible. In contrast, missing data prohibit the evaluation of the time-series operators that involve missing observations. Thus, the above example is not equivalent to the following one:

```
. use http://www.stata-press.com/data/r8/abdata

. xtabond n l(0/1).w l(0/2).(k ys) yr1980-yr1984 if year!=1980, lags(2) robust
note: yr1980 dropped due to collinearity
```

```
Arellano-Bond dynamic panel-data estimation      Number of obs       =       473
Group variable (i): id                           Number of groups    =       140

                                                 Wald chi2(14)       =    463.79

Time variable (t): year                          Obs per group: min =         3
                                                                avg = 3.378571
                                                                max =         5
```

One-step results

n		Coef.	Robust Std. Err.	z	P>|z|	[95% Conf. Interval]	
n							
	LD	.7187143	.1304986	5.51	0.000	.4629417	.9744868
	L2D	-.0938608	.0583746	-1.61	0.108	-.2082729	.0205512
w							
	D1	-.6599329	.1753661	-3.76	0.000	-1.003644	-.3162217
	LD	.4654663	.1657696	2.81	0.005	.1405638	.7903688
k							
	D1	.3894823	.0731561	5.32	0.000	.2460989	.5328657
	LD	-.1122138	.0892247	-1.26	0.209	-.287091	.0626634
	L2D	-.0228185	.037552	-0.61	0.543	-.0964191	.050782
ys							
	D1	.4683059	.1830265	2.56	0.011	.1095805	.8270313
	LD	-.8632798	.2205727	-3.91	0.000	-1.295594	-.4309653
	L2D	.0991026	.1436945	0.69	0.490	-.1825336	.3807387
yr1981							
	D1	-.0655563	.0207481	-3.16	0.002	-.1062219	-.0248907
yr1982							
	D1	-.1092257	.0329365	-3.32	0.001	-.1737801	-.0446714
yr1983							
	D1	-.1268257	.0445178	-2.85	0.004	-.2140791	-.0395724
yr1984							
	D1	-.1249563	.0531627	-2.35	0.019	-.2291533	-.0207593
_cons		.0146263	.0107159	1.36	0.172	-.0063766	.0356291

```
Arellano-Bond test that average autocovariance in residuals of order 1 is 0:
         H0: no autocorrelation   z =  -3.85   Pr > z = 0.0001
Arellano-Bond test that average autocovariance in residuals of order 2 is 0:
         H0: no autocorrelation   z =   1.37   Pr > z = 0.1707
```

In this case, the year 1980 is dropped from the sample, but when the value of a variable from 1980 is required because a lag or difference is required, the 1980 value is used.

◁

(*Continued on next page*)

Saved Results

xtabond saves in e():

Scalars

e(N)	number of observations	e(F)	model F statistic (small only)
e(N_g)	number of groups	e(F_p)	p-value from model F (small only)
e(df_m)	model degrees of freedom	e(F_df)	restrictions in model F (small only)
e(g_max)	largest group size	e(df_r)	denominator df in F (small only)
e(g_min)	smallest group size	e(chi2)	model $\chi 2$ statistic
e(g_avg)	average group size	e(chi2_p)	p-value from model $\chi 2$
e(t_max)	maximum time in sample	e(chi2_df)	restrictions in model $\chi 2$
e(t_min)	minimum time in sample	e(arm#)	test for autocorrelation of order #
e(n_lags)	number of lags of e(depvar)	e(sig2)	estimate of σ_ϵ^2
e(sargon)	Sargon test statistic	e(artests)	number of AR tests performed
e(sar_df)	degrees of freedom for sargon	e(zcols)	number of columns in instrument matrix Z_i

Macros

e(cmd)	xtabond	e(tvar)	time variable
e(depvar)	name of dependent variable	e(vcetype)	covariance estimation method
e(inst_l)	additional level instruments	e(chi2type)	Wald; type of model χ^2 test
e(twostep)	twostep, if specified	e(robust)	robust, if specified
e(ivar)	variable denoting groups	e(predict)	program used to implement predict

Matrices

e(b)	coefficient vector	e(V)	variance–covariance matrix of the estimators

Functions

e(sample)	marks estimation sample

Methods and Formulas

xtabond is implemented as an ado-file.

Consider dynamic panel-data models of the form

$$y_{it} = \sum_{j=1}^{p} \alpha_j y_{i,t-j} + \mathbf{x}_{it}\boldsymbol{\beta}_1 + \mathbf{w}_{it}\boldsymbol{\beta}_2 + \nu_i + \epsilon_{it}$$

where the variables are as defined as in (1).

Note that $\mathbf{x}$ and $\mathbf{w}$ may contain lagged independent variables and time dummies.

Let $X_{it} = (y_{i,t-1}, y_{i,t-2}, \ldots, y_{i,t-p}, \mathbf{x}_{it}, \mathbf{w}_{it})$ be the $1 \times K$ vector of covariates for i at time t, where $K = p + k_1 + k_2$.

Now, rewrite this relationship as a set of N equations for each individual:

$$y_i = X_i \boldsymbol{\delta} + \nu_i \iota_i + \epsilon_i$$

Simplifying the notation, assume that there are no more than p lags of any of the covariates. Then, for each i, y_i, ι_i, and ϵ_i are all $(T_i - p) \times 1$ vectors; y_i is just the stacked values of y_{it} for person i, ι_i is a vector of ones, and ϵ_i contains the stacked values of ϵ_{it} for person i. The matrix X_i contains the p lags of y_{it}, the values of $\mathbf{x}_{it}$, and the $\mathbf{w}_{it}$. And, $\boldsymbol{\delta}$ is the $K \times 1$ vector of coefficients.

Define the first-differenced versions as

$$y_i^* = \begin{pmatrix} y_{i,t_{i0}+1+p} - y_{i,t_{i0}+p} \\ y_{i,t_{i0}+2+p} - y_{i,t_{i0}+1+p} \\ \vdots \\ y_{iT_i} - y_{i,T_i-1} \end{pmatrix}$$

$$X_i^* = \begin{pmatrix} X_{i,t_{i0}+1+p} - X_{i,t_{i0}+p} \\ X_{i,t_{i0}+2+p} - X_{i,t_{i0}+1+p} \\ \vdots \\ X_{iT_i} - X_{i,T_i-1} \end{pmatrix}$$

$$\epsilon_i^* = \begin{pmatrix} \epsilon_{i,t_{i0}+1+p} - \epsilon_{i,t_{i0}+p} \\ \epsilon_{i,t_{i0}+2+p} - \epsilon_{i,t_{i0}+1+p} \\ \vdots \\ \epsilon_{iT_i} - \epsilon_{i,T_i-1} \end{pmatrix}$$

where t_{i0} is the time period of the first nonmissing observation for person i.

The most difficult part of these estimators is defining and implementing the matrix of instruments for each i, Z_i. Begin by considering a simple balanced panel example in which our model is

$$y_{it} = y_{i,t-1}\alpha_1 + y_{i,t-2}\alpha_2 + \mathbf{x}_{it}\boldsymbol{\beta} + \nu_i + \epsilon_{it}$$

Note that there are no predetermined variables. Further simplify the situation by assuming that the data come from a balanced panel in which there are no missing values. After first differencing the equation, we have

$$\Delta y_{it} = \Delta y_{i,t-1}\alpha_1 + \Delta y_{i,t-2}\alpha_2 + \Delta \mathbf{x}_{it}\boldsymbol{\beta} + \Delta \epsilon_{it}$$

The first three observations are lost to lags and differencing. Since $\mathbf{x}_{it}$ contains only strictly exogenous covariates, Δx_{it} will serve as its own instrument in estimating the first-differenced equation. Assuming that ϵ_{it} are not autocorrelated, for each i at $t = 4$, y_{i1} and y_{i2} are valid for the lagged variables. Similarly, at $t = 5$, y_{i1}, y_{i2}, and y_{i3} are valid instruments. Continuing in this fashion, we obtain an instrument matrix with one row for each time period that we are instrumenting:

$$Z_i = \begin{pmatrix} y_{i2} & y_{i3} & 0 & 0 & 0 & \cdots & 0 & 0 & 0 & \Delta\mathbf{x}_{i5} \\ 0 & 0 & y_{i1} & y_{i2} & y_{i3} & \cdots & 0 & 0 & 0 & \Delta\mathbf{x}_{i6} \\ \vdots & \vdots & \vdots & \vdots & \ddots & \vdots & \vdots & \vdots & \vdots & \vdots \\ 0 & 0 & 0 & 0 & \cdots & 0 & y_{i2} & \cdots & y_{i,T-2} & \Delta\mathbf{x}_{iT} \end{pmatrix}$$

Since $p = 2$, note that Z_i has $T - p - 1$ rows and $\sum_{m=p}^{T-2} m + k_1$ columns, where k_1 is the number of variables in x.

The extension to other lag structures with complete data is immediate. Unbalanced data and missing observations are handled by dropping the rows for which there are no data, and filling in zeros in columns where missing data would be required. For instance, suppose that for some i, the $t = 1$ observation was missing, but that this observation was not missing for some other panels. Our instrument matrix would then be

$$Z_i = \begin{pmatrix} 0 & 0 & 0 & y_{i1} & y_{i2} & 0 & 0 & 0 & 0 & \ldots & 0 & 0 & 0 & \Delta\mathbf{x}_{i4} \\ 0 & 0 & 0 & 0 & 0 & y_{i1} & y_{i2} & y_{i3} & 0 & \ldots & 0 & 0 & 0 & \Delta\mathbf{x}_{i5} \\ \vdots & \vdots & \vdots & \vdots & \vdots & \vdots & \vdots & \vdots & \vdots & \ddots & \vdots & \vdots & \vdots & \vdots \\ 0 & 0 & 0 & 0 & 0 & 0 & 0 & 0 & \ldots & y_{i1} & y_{i2} & \ldots & y_{iT-2} & \Delta\mathbf{x}_{iT} \end{pmatrix}$$

Note that Z_i has $T_i - p - 1$ rows and $\sum_{m=p}^{\tau-2} m + k_1$ columns, where $\tau = \max_i \tau_i$ and τ_i is the number of nonmissing observations in panel i.

Endogenous variables are treated similarly to the lagged dependent variables, and levels lagged 2 or more periods are valid instruments. For predetermined variables, levels lagged 1 or more periods are valid instruments. See Arellano and Bond (1991, 290) for an example.

Note that the number of columns in Z_i can grow very quickly for moderately long panels or for models with several predetermined variables.

Let H_i be the $(T_i - p - 1) \times (T_i - p - 1)$ covariance matrix of the differenced idiosyncratic errors; i.e.,

$$H_i = E[\epsilon_i^* \epsilon_i^{*'}] = \begin{pmatrix} 2 & -1 & 0 & \ldots & 0 & 0 \\ -1 & 2 & -1 & \ldots & 0 & 0 \\ \vdots & \vdots & \vdots & \ddots & \vdots & \vdots \\ 0 & 0 & 0 & \ldots & 2 & -1 \\ 0 & 0 & 0 & \ldots & -1 & 2 \end{pmatrix}$$

Then, for some instrument matrix Z_i, the one-step Arellano–Bond estimator of δ, $\widehat{\delta}_1$, is given by

$$\widehat{\delta}_1 = Q_1^{-1} \left(\sum_{i=1}^{N} X_i^{*'} Z_i \right) A_1 \left(\sum_{i=1}^{N} Z_i' y_i^* \right)$$

where

$$Q_1 = \left(\sum_{i=1}^{N} X_i^{*'} Z_i \right) A_1 \left(\sum_{i=1}^{N} Z_i' X_i^{*'} \right)$$

and

$$A_1 = \sum_{i=1}^{N} Z_i' H_i Z_i$$

Using $\widehat{\delta}_1$, the one-step residuals for i are

$$\widehat{\epsilon}_i^* = y_i^* - X_i^* \widehat{\delta}_1$$

Assuming homoskedasticity, the variance–covariance estimator of the parameter estimator $\widehat{\delta}_1$ is

$$\widehat{V}_1 = \widehat{\sigma}_1^2 Q_1^{-1}$$

where

$$\widehat{\sigma}_1^2 = \frac{1}{NT - K} \sum_{i=1}^{N} (\widehat{\epsilon}_i^{*'} \widehat{\epsilon}_i)^*$$

and $NT = \sum_{i=1}^{N} T_i - p - 1$.

The robust estimator is given by

$$\widehat{V}_{1r} = Q_1^{-1} \left(\sum_{i=1}^{N} X_i^{*'} Z_i \right) A_1 A_2^{-1} A_1 \left(\sum_{i=1}^{N} Z_i' X_i^* \right) Q_1^{-1}$$

where

$$A_2 = \sum_{i=1}^{N} Z_i' G_i Z_i$$

and

$$G_i = \widehat{\epsilon}_i^* \widehat{\epsilon}_i^{*'}$$

The two-step estimator of δ, $\widehat{\delta}_2$, is given by

$$\widehat{\delta}_2 = Q_2^{-1} \left(\sum_{i=1}^{N} X_i^{*'} Z_i \right) A_2 \left(\sum_{i=1}^{N} Z_i' y_i^* \right)$$

where

$$Q_2 = \left(\sum_{i=1}^{N} X_i^{*'} Z_i \right) A_2 \left(\sum_{i=1}^{N} Z_i' X_i^{*'} \right)$$

The two-step VCE is

$$\widehat{V}_2 = Q_2^{-1}$$

For the one-step, homoskedastic case, the test for autocorrelation of order m in the differenced residuals $\widehat{\epsilon}_i^*$ is given by

$$AR_m = \frac{\sum_{i=1}^{N} (\widehat{\epsilon}_{mi}^{*'} \widehat{\epsilon}_i^*)}{B_1^{1/2}}$$

where

$$\widehat{\epsilon}_{mi}^* = L_m(\widehat{\epsilon}_i^*)$$

and L_m is the m-order lag operator, and

$$B_1 = \sum_{i=1}^{N} \widehat{\epsilon}_{mi}^{*'} H_i \widehat{\epsilon}_{mi}^* - 2 \left(\sum_{i=1}^{N} \widehat{\epsilon}_{mi}^{*'} X_i^* \right) Q_1^{-1} \left(\sum_{i=1}^{N} X_i^{*'} Z_i \right) A_1 \left(\sum_{i=1}^{N} Z_i' H_i \widehat{\epsilon}_{mi}^* \right)$$
$$+ \left(\sum_{i=1}^{N} \widehat{\epsilon}_{mi}^{*'} X_i^* \right) \widehat{V}_1 \left(\sum_{i=1}^{N} X_i^{*'} \widehat{\epsilon}_{mi}^* \right)$$

For the one-step, robust case, the test becomes

$$AR_m = \frac{\sum_{i=1}^{N} (\widehat{\epsilon}_{mi}^{*'} \widehat{\epsilon}_i^*)}{B_{1r}^{1/2}}$$

where

$$B_{1r} = \sum_{i=1}^{N} \widehat{\epsilon}_{mi}^{*'} G_i \widehat{\epsilon}_{mi}^{*} - 2\left(\sum_{i=1}^{N} \widehat{\epsilon}_{mi}^{*'} X_i^{*}\right) Q_1^{-1} \left(\sum_{i=1}^{N} X_i^{*'} Z_i\right) A_2 \left(\sum_{i=1}^{N} Z_i' G_i \widehat{\epsilon}_{mi}^{*}\right)$$
$$+ \left(\sum_{i=1}^{N} \widehat{\epsilon}_{mi}^{*'} X_i^{*}\right) \widehat{V}_{1r} \left(\sum_{i=1}^{N} X_i^{*'} \widehat{\epsilon}_{mi}^{*}\right)$$

For the two-step case, the test is

$$AR_m = \frac{\sum_{i=1}^{N} \left(\widehat{\widehat{\epsilon}}_{mi}^{*'} \widehat{\widehat{\epsilon}}_i^{*}\right)}{B_2^{1/2}}$$

where

$$B_2 = \sum_{i=1}^{N} \widehat{\widehat{\epsilon}}_{mi}^{*'} G_{2i} \widehat{\widehat{\epsilon}}_{mi}^{*} - 2\left(\sum_{i=1}^{N} \widehat{\widehat{\epsilon}}_{mi}^{*'} X_i^{*}\right) Q_2^{-1} \left(\sum_{i=1}^{N} X_i^{*'} Z_i\right) A_2 \left(\sum_{i=1}^{N} Z_i' G_{2i} \widehat{\widehat{\epsilon}}_{mi}^{*}\right)$$
$$+ \left(\sum_{i=1}^{N} \widehat{\widehat{\epsilon}}_{mi}^{*'} X_i^{*}\right) \widehat{V}_2 \left(\sum_{i=1}^{N} X_i^{*'} \widehat{\widehat{\epsilon}}_{mi}^{*}\right)$$

where

$$\widehat{\widehat{\epsilon}}_i^{*} = y_i^{*} - X_i^{*} \widehat{\delta}_2$$

and

$$\widehat{\widehat{\epsilon}}_{mi}^{*} = L_m(\widehat{\widehat{\epsilon}}_i^{*})$$

and

$$G_{2i} = \widehat{\widehat{\epsilon}}_i^{*} \widehat{\widehat{\epsilon}}_i^{*'}$$

The Sargan test statistic for the one-step model is

$$S_1 = \left(\sum_{i=1}^{N} \widehat{\epsilon}_i^{*'} Z_i\right) A_1 \left(\sum_{i=1}^{N} Z_i' \widehat{\epsilon}_i^{*}\right) \widehat{\sigma}_\epsilon^2$$

The Sargan test statistic for the two-step model is

$$S_2 = \left(\sum_{i=1}^{N} \widehat{\widehat{\epsilon}}_i^{*'} Z_i\right) A_2 \left(\sum_{i=1}^{N} Z_i' \widehat{\widehat{\epsilon}}_i^{*}\right)$$

References

Anderson, T. W. and C. Hsiao. 1981. Estimation of dynamic models with error components. *Journal of the American Statistical Association* 76: 598–606.

——. 1982. Formulation and estimation of dynamic models using panel data. *Journal of Econometrics* 18: 47–82.

Arellano, M. and S. Bond. 1991. Some tests of specification for panel data: Monte Carlo evidence and an application to employment equations. *The Review of Economic Studies* 58: 277–297.

Baltagi, B. H. 2001. *Econometric Analysis of Panel Data*. 2d ed. New York: John Wiley & Sons.

Hansen, L. P. 1982. Large sample properties of generalized method of moments estimators. *Econometrica* 50: 1029–1054.

Kiviet, J. 1995. On bias, inconsistency, and efficiency of various estimators in dynamic panel data models. *Journal of Econometrics* 68: 53–78.

Layard, R. and S. J. Nickell. 1986. Unemployment in Britain. *Economica* 53: 5121–5169.

Also See

Complementary:	[XT] **xtdata**, [XT] **xtdes**, [XT] **xtsum**, [XT] **xttab**,
	[R] **adjust**, [R] **lincom**, [R] **mfx**, [R] **nlcom**, [R] **predict**,
	[R] **predictnl**, [R] **test**, [R] **testnl**, [R] **vce**,
	[TS] **tsset**
Related:	[XT] **xtgee**, [XT] **xthtaylor**, [XT] **xtintreg**, [XT] **xtivreg**, [XT] **xtreg**,
	[XT] **xtregar**, [XT] **xttobit**
Background:	[U] **16.5 Accessing coefficients and standard errors**,
	[U] **23 Estimation and post-estimation commands**,
	[U] **23.14 Obtaining robust variance estimates**,
	[XT] **xt**

Title

> **xtcloglog** — Random-effects and population-averaged cloglog models

Syntax

Random-effects model

> xtcloglog *depvar* [*varlist*] [*weight*] [if *exp*] [in *range*] [, re i(*varname*) quad(*#*)
>
> noconstant noskip level(*#*) offset(*varname*) nolog *maximize_options*]

Population-averaged model

> xtcloglog *depvar* [*varlist*] [*weight*] [if *exp*] [in *range*] , pa [i(*varname*) robust
>
> noconstant level(*#*) offset(*varname*) nolog *xtgee_options maximize_options*]

by ... : may be used with xtcloglog; see [R] **by**.

iweights, fweights, and pweights are allowed for the population-averaged model and iweights are allowed for the random-effects model; see [U] **14.1.6 weight**. Note that weights must be constant within panels.

xtcloglog shares the features of all estimation commands; see [U] **23 Estimation and post-estimation commands**.

Syntax for predict

Random-effects model

> predict [*type*] *newvarname* [if *exp*] [in *range*] [, [xb | pu0 | stdp]
>
> nooffset]

Population-averaged model

> predict [*type*] *newvarname* [if *exp*] [in *range*] [, [mu | rate | xb | stdp]
>
> nooffset]

These statistics are available both in and out of sample; type predict ... if e(sample) ... if wanted only for the estimation sample.

Description

xtcloglog fits population-averaged and random-effects complementary log-log (cloglog) models. There is no command for a conditional fixed-effects model, as there does not exist a sufficient statistic allowing the fixed effects to be conditioned out of the likelihood. Unconditional fixed-effects cloglog models may be fitted with the cloglog command with indicator variables for the panels. The appropriate indicator variables can be generated using tabulate or xi. However, unconditional fixed-effects estimates are biased.

By default, the population-averaged model is an equal-correlation model; that is, xtcloglog, pa assumes corr(exchangeable). See [XT] **xtgee** for details on how to fit other population-averaged models.

Note: xtcloglog, re, the default, is slow since it is calculated by quadrature; see *Methods and Formulas*. Computation time is roughly proportional to the number of points used for the quadrature. The default is quad(12). Simulations indicate that increasing it does not appreciably change the estimates for the coefficients or their standard errors. See [XT] **quadchk**.

See [R] **logistic** for a list of related estimation commands.

Options

re requests the random-effects estimator. re is the default if neither re nor pa is specified.

pa requests the population-averaged estimator.

i(*varname*) specifies the variable name that contains the unit to which the observation belongs. You can specify the i() option the first time you estimate, or you can use the iis command to set i() beforehand. Note that it is not necessary to specify i() if the data have been previously tsset, or if iis has been previously specified—in these cases, the group variable is taken from the previous setting. See [XT] **xt**.

quad(*#*) specifies the number of points to use in the quadrature approximation of the integral. The default is quad(12).

robust specifies that the Huber/White/sandwich estimator of variance is to be used in place of the IRLS variance estimator; see [XT] **xtgee**. This alternative produces valid standard errors even if the correlations within group are not as hypothesized by the specified correlation structure. It does, however, require that the model correctly specifies the mean. As such, the resulting standard errors are labeled "semi-robust" instead of "robust". Note that although there is no cluster() option, results are as if there were a cluster() option and you specified clustering on i().

noconstant suppresses the constant term (intercept) in the model.

noskip specifies that a full maximum-likelihood model with only a constant for the regression equation be fitted. This model is not displayed, but is used as the base model to compute a likelihood-ratio test for the model test statistic displayed in the estimation header. By default, the overall model test statistic is an asymptotically equivalent Wald test of all the parameters in the regression equation being zero (except the constant). For many models, this option can substantially increase estimation time.

level(*#*) specifies the confidence level, in percent, for confidence intervals. The default is level(95) or as set by set level; see [U] **23.6 Specifying the width of confidence intervals**.

offset(*varname*) specifies that *varname* is to be included in the model with its coefficient constrained to be 1.

nolog suppresses the iteration log.

xtgee_options specifies any other options allowed by xtgee for family(binomial) link(cloglog) such as corr(); see [XT] **xtgee**.

maximize_options control the maximization process; see [R] **maximize**. Use the trace option to view parameter convergence. Use the ltol(*#*) option to relax the convergence criterion; the default is 1e−6 during specification searches.

Options for predict

xb calculates the linear prediction. This is the default for the random-effects model.

pu0 calculates the probability of a positive outcome, assuming that the random effect for that observation's panel is zero ($\nu = 0$). Note that this may not be similar to the proportion of observed outcomes in the group.

stdp calculates the standard error of the linear prediction.

mu and rate both calculate the predicted probability of *depvar*. mu takes into account the offset(). rate ignores those adjustments. mu and rate are equivalent if you did not specify offset(). mu is the default for the population-averaged model.

nooffset is relevant only if you specified offset(*varname*) for xtcloglog. It modifies the calculations made by predict so that they ignore the offset variable; the linear prediction is treated as $\mathbf{x}_{it}\boldsymbol{\beta}$ rather than $\mathbf{x}_{it}\boldsymbol{\beta} + \text{offset}_{it}$.

Remarks

xtcloglog, pa is a convenience command if you want the population-averaged model. Typing

        . xtcloglog ..., pa ...

is equivalent to typing

        . xtgee ..., ... family(binomial) link(cloglog) corr(exchangeable)

Thus, also see [XT] **xtgee** for information about xtcloglog.

By default, or when re is specified, xtcloglog fits via maximum-likelihood the random-effects model

$$\Pr(y_{it} \neq 0 | \mathbf{x}_{it}) = P(\mathbf{x}_{it}\boldsymbol{\beta} + \nu_i)$$

for $i = 1, \ldots, n$ panels, $t = 1, \ldots, n_i$, ν_i are iid $N(0, \sigma_\nu^2)$, and $P(z) = 1 - \exp\{-\exp(z)\}$.

Underlying this model is the variance components model

$$y_{it} \neq 0 \iff \mathbf{x}_{it}\boldsymbol{\beta} + \nu_i + \epsilon_{it} > 0$$

where ϵ_{it} are iid Extreme-value (Gumbel) distributed with mean equal to Euler's constant and variance $\sigma_\epsilon^2 = \pi^2/6$, independently of ν_i. The nonsymmetric error distribution is an alternative to logit and probit analysis, and this model is typically used when the positive (or negative) outcome is rare.

▷ Example

You are studying unionization of women in the United States and are using the union dataset; see [XT] **xt**. You wish to fit a random-effects model of union membership:

(*Continued on next page*)

```
. use http://www.stata-press.com/data/r8/union
(NLS Women 14-24 in 1968)

. xtcloglog union age grade not_smsa south southXt, i(id) nolog
```

Random-effects complementary log-log model	Number of obs =	26200
Group variable (i): idcode	Number of groups =	4434
Random effects u_i ~ Gaussian	Obs per group: min =	1
	avg =	5.9
	max =	12
	Wald chi2(5) =	221.84
Log likelihood = -10559.721	Prob > chi2 =	0.0000

union	Coef.	Std. Err.	z	P>|z|	[95% Conf. Interval]	
age	.011177	.0032236	3.47	0.001	.0048588	.0174952
grade	.0577535	.012636	4.57	0.000	.0329875	.0825195
not_smsa	-.2122413	.0613769	-3.46	0.001	-.3325379	-.0919447
south	-.8724097	.0851594	-10.24	0.000	-1.039319	-.7055004
southXt	.0173595	.0059679	2.91	0.004	.0056626	.0290564
_cons	-3.066854	.1816358	-16.88	0.000	-3.422853	-2.710854
/lnsig2u	1.158696	.0392705			1.081727	1.235665
sigma_u	1.784874	.0350464			1.717489	1.854903
rho	.659484	.0088188			.6419935	.6765506

```
Likelihood-ratio test of rho=0: chibar2(01) =  5968.96 Prob >= chibar2 = 0.000
```

The output includes the additional panel-level variance component. This is parameterized as the log of the standard deviation $\ln \sigma_\nu$ (labeled `lnsig2u` in the output). The standard deviation σ_ν is also included in the output, labeled `sigma_u`, together with ρ (labeled `rho`),

$$\rho = \frac{\sigma_\nu^2}{\sigma_\nu^2 + \sigma_\epsilon^2}$$

which is the proportion of the total variance contributed by the panel-level variance component.

When `rho` is zero, the panel-level variance component is unimportant and the panel estimator is no different from the pooled estimator (`cloglog`). A likelihood-ratio test of this is included at the bottom of the output. This test formally compares the pooled estimator with the panel estimator.

As an alternative to the random-effects specification, you might want to fit an equal-correlation population-averaged cloglog model by typing

(Continued on next page)

```
. xtcloglog union age grade not_smsa south southXt, i(id) pa
Iteration 1: tolerance = .06580809
Iteration 2: tolerance = .00606963
Iteration 3: tolerance = .00032265
Iteration 4: tolerance = .00001658
Iteration 5: tolerance = 8.864e-07
```

```
GEE population-averaged model              Number of obs      =        26200
Group variable:                    idcode  Number of groups   =         4434
Link:                             cloglog  Obs per group: min =            1
Family:                          binomial                 avg =          5.9
Correlation:                 exchangeable                 max =           12
                                           Wald chi2(5)       =       232.44
Scale parameter:                        1  Prob > chi2        =       0.0000
```

union	Coef.	Std. Err.	z	P>|z|	[95% Conf.	Interval]
age	.0045777	.0021754	2.10	0.035	.0003139	.0088415
grade	.0544267	.0095097	5.72	0.000	.035788	.0730654
not_smsa	-.1051731	.0430512	-2.44	0.015	-.189552	-.0207943
south	-.6578891	.061857	-10.64	0.000	-.7791266	-.5366515
southXt	.0142329	.004133	3.44	0.001	.0061325	.0223334
_cons	-2.074687	.1358008	-15.28	0.000	-2.340851	-1.808522

◁

▷ Example

In [R] **cloglog**, we showed the above results and compared them with cloglog, robust cluster(). xtcloglog with the pa option allows a robust option (the random-effects estimator does not allow the robust specification), and so we can obtain the population-averaged cloglog estimator with the robust variance calculation by typing

```
. xtcloglog union age grade not_smsa south southXt, i(id) pa robust nolog
GEE population-averaged model              Number of obs      =        26200
Group variable:                    idcode  Number of groups   =         4434
Link:                             cloglog  Obs per group: min =            1
Family:                          binomial                 avg =          5.9
Correlation:                 exchangeable                 max =           12
                                           Wald chi2(5)       =       153.64
Scale parameter:                        1  Prob > chi2        =       0.0000
```

            (standard errors adjusted for clustering on idcode)

union	Coef.	Semi-robust Std. Err.	z	P>|z|	[95% Conf.	Interval]
age	.0045777	.003261	1.40	0.160	-.0018138	.0109692
grade	.0544267	.0117512	4.63	0.000	.0313948	.0774585
not_smsa	-.1051731	.0548342	-1.92	0.055	-.2126462	.0022999
south	-.6578891	.0793619	-8.29	0.000	-.8134355	-.5023427
southXt	.0142329	.005975	2.38	0.017	.0025221	.0259438
_cons	-2.074687	.1770236	-11.72	0.000	-2.421647	-1.727727

These standard errors are similar to those shown for cloglog, robust cluster() in [R] **cloglog**.

◁

Saved Results

xtcloglog, re saves in e():

Scalars

e(N)	number of observations	e(g_avg)	average group size
e(N_g)	number of groups	e(chi2)	χ^2
e(df_m)	model degrees of freedom	e(chi2_c)	χ^2 for comparison test
e(ll)	log likelihood	e(rho)	ρ
e(ll_0)	log likelihood, constant-only model	e(sigma_u)	panel-level standard deviation
e(ll_c)	log likelihood, comparison model	e(N_cd)	number of completely determined obs.
e(g_max)	largest group size	e(n_quad)	number of quadrature points
e(g_min)	smallest group size		

Macros

e(cmd)	xtcloglog	e(chi2type)	Wald or LR; type of model χ^2 test
e(depvar)	name of dependent variable	e(chi2_ct)	Wald or LR; type of model χ^2 test
e(title)	title in estimation output		corresponding to e(chi2_c)
e(ivar)	variable denoting groups	e(distrib)	Gaussian; the distribution of the
e(wtype)	weight type		random effect
e(wexp)	weight expression	e(crittype)	optimization criterion
e(offset)	offset	e(predict)	program used to implement predict

Matrices

e(b)	coefficient vector	e(V)	variance–covariance matrix of the estimators

Functions

e(sample)	marks estimation sample

xtcloglog, pa saves in e():

Scalars

e(N)	number of observations	e(deviance)	deviance
e(N_g)	number of groups	e(chi2_dev)	χ^2 test of deviance
e(df_m)	model degrees of freedom	e(dispers)	deviance dispersion
e(g_max)	largest group size	e(chi2_dis)	χ^2 test of deviance dispersion
e(g_min)	smallest group size	e(tol)	target tolerance
e(g_avg)	average group size	e(dif)	achieved tolerance
e(chi2)	χ^2	e(phi)	scale parameter
e(df_pear)	degrees of freedom for Pearson χ^2		

Macros

e(cmd)	xtgee	e(scale)	x2, dev, phi, or #; scale parameter
e(cmd2)	xtcloglog	e(ivar)	variable denoting groups
e(depvar)	name of dependent variable	e(vcetype)	covariance estimation method
e(family)	binomial	e(chi2type)	Wald; type of model χ^2 test
e(link)	cloglog; link function	e(offset)	offset
e(corr)	correlation structure	e(predict)	program used to implement predict
e(crittype)	optimization criterion		

Matrices

e(b)	coefficient vector	e(R)	estimated working correlation matrix
e(V)	variance–covariance matrix of the estimators		

Functions

e(sample)	marks estimation sample

Methods and Formulas

xtcloglog is implemented as an ado-file.

xtcloglog reports the population-averaged results obtained by using xtgee, family(binomial) link(cloglog) to obtain estimates.

Assuming a normal distribution, $N(0, \sigma_\nu^2)$, for the random effects ν_i, we have that

$$\Pr(y_{i1}, \ldots, y_{in_i} | \mathbf{x}_{i1}, \ldots, \mathbf{x}_{in_i}) = \int_{-\infty}^{\infty} \frac{e^{-\nu_i^2/2\sigma_\nu^2}}{\sqrt{2\pi}\sigma_\nu} \left\{ \prod_{t=1}^{n_i} F(y_{it}, \mathbf{x}_{it}\boldsymbol{\beta} + \nu_i) \right\} d\nu_i$$

where

$$F(y, z) = \begin{cases} 1 - \exp\left\{ -\exp(z) \right\} & \text{if } y \neq 0 \\ \exp\left\{ -\exp(z) \right\} & \text{otherwise} \end{cases}$$

We can approximate the integral with M-point Gauss–Hermite quadrature

$$\int_{-\infty}^{\infty} e^{-x^2} g(x)dx \approx \sum_{m=1}^{M} w_m^* g(a_m^*)$$

where the w_m^* denote the quadrature weights and the a_m^* denote the quadrature abscissas. The log-likelihood L, where $\tau = \sigma_\nu^2/(\sigma_\nu^2 + 1)$, is then calculated using the quadrature

$$L = \sum_{i=1}^{n} w_i \log\left\{ \Pr(y_{i1}, \ldots, y_{in_i} | \mathbf{x}_{i1}, \ldots, \mathbf{x}_{in_i}) \right\}$$

$$\approx \sum_{i=1}^{n} w_i \log\left[\frac{1}{\sqrt{\pi}} \sum_{m=1}^{M} w_m^* \prod_{t=1}^{n_i} F\left\{ y_{it}, \mathbf{x}_{it}\boldsymbol{\beta} + a_m^* \left(\frac{2\tau}{1-\tau} \right)^{1/2} \right\} \right]$$

where w_i is the user-specified weight for panel i; if no weights are specified, $w_i = 1$.

The quadrature formula requires that the integrated function be well-approximated by a polynomial. As the number of time periods becomes large (as panel size gets large),

$$\prod_{t=1}^{n_i} F(y_{it}, \mathbf{x}_{it}\boldsymbol{\beta} + \nu_i)$$

is no longer well-approximated by a polynomial. As a general rule of thumb, you should use this quadrature approach only for small to moderate panel sizes (based on simulations, 50 is a reasonably safe upper bound). However, if the data really come from random-effects cloglog and τ is not too large (less than, say, .3), then the panel size could be 500 and the quadrature approximation would still be fine. If the data are not random-effects cloglog or τ is large (bigger than, say, .7), then the quadrature approximation may be poor for panel sizes larger than 10. The quadchk command should be used to investigate the applicability of the numeric technique used in this command.

References

Liang, K.-Y. and S. L. Zeger. 1986. Longitudinal data analysis using generalized linear models. *Biometrika* 73: 13–22.

Neuhaus, J. M. 1992. Statistical methods for longitudinal and clustered designs with binary responses. *Statistical Methods in Medical Research* 1: 249–273.

Neuhaus, J. M., J. D. Kalbfleisch, and W. W. Hauck. 1991. A comparison of cluster-specific and population-averaged approaches for analyzing correlated binary data. *International Statistical Review* 59: 25–35.

Pendergast, J. F., S. J. Gange, M. A. Newton, M. J. Lindstrom, M. Palta, and M. R. Fisher. 1996. A survey of methods for analyzing clustered binary response data. *International Statistical Review* 64: 89–118.

Also See

Complementary:	[XT] **quadchk**, [XT] **xtdata**, [XT] **xtdes**, [XT] **xtsum**, [XT] **xttab**,
	[R] **adjust**, [R] **lincom**, [R] **mfx**, [R] **nlcom**, [R] **predict**, [R] **predictnl**,
	[R] **test**, [R] **testnl**, [R] **vce**
Related:	[XT] **xtgee**, [XT] **xtlogit**, [XT] **xtprobit**,
	[R] **cloglog**
Background:	[U] **16.5 Accessing coefficients and standard errors**,
	[U] **23 Estimation and post-estimation commands**,
	[U] **23.14 Obtaining robust variance estimates**,
	[XT] **xt**

Title

xtdata — Faster specification searches with xt data

Syntax

xtdata [*varlist*] [if *exp*] [in *range*] [, re be fe r̲atio(*#*) clear i(*varname*) nodouble]

Description

xtdata produces a converted dataset of the variables specified, or, if *varlist* is not specified, all the variables in the data. Once converted, Stata's ordinary regress command may be used to perform specification searches more quickly than use of xtreg; see [R] **regress** and [XT] **xtreg**. In the case of xtdata, re, a variable named constant is also created. When using regress after xtdata, re, specify noconstant and include constant in the regression. After xtdata, be and xtdata, fe, you need not include constant or specify regress's noconstant option.

Options

re specifies that the data are to be converted into a form suitable for random-effects estimation. re is the default if none of be, fe, or re are specified. ratio() must also be specified.

be specifies that the data are to be converted into a form suitable for between estimation.

fe specifies that the data are to be converted into a form suitable for fixed-effects (within) estimation.

ratio(*#*), used with xtdata, re only, specifies the ratio $\sigma_\nu/\sigma_\epsilon$, the ratio of the random effect to the pure residual. Note that this is the ratio of the standard deviations, not the variances.

clear specifies that the data may be converted even though the dataset has changed since it was last saved on disk.

i(*varname*) specifies the variable name corresponding to i in x_{it}. You can specify the i() option the first time you estimate, or you can use the iis command to set i() beforehand. Note that it is not necessary to specify i() if the data have been previously tsset, or if iis has been previously specified—in these cases, the group variable is taken from the previous setting. See [XT] **xt**.

nodouble specifies that transformed variables should, in general, keep their original types if possible. The default is to recast variables to double.

Remember that xtdata transforms variables to be differences from group means, pseudo-differences from group means, or group means. Specifying nodouble will decrease the size of the resulting dataset, but may introduce round-off errors in these calculations.

Remarks

If you have not read [XT] **xt** and [XT] **xtreg**, please do so.

The formal estimation commands of xtreg—see [XT] **xtreg**—are not instant, especially with large datasets. Equations (2), (3), and (4) of [XT] **xtreg** provide a description of the data necessary to fit each of the models with OLS. The idea here is to transform the data once to the appropriate form, and then use regress to fit such models more quickly.

▷ Example

Please see the example in [XT] **xtreg** demonstrating between-effects regression. An alternative way to estimate the between equation is to convert the data in memory into the between data:

```
. use http://www.stata-press.com/data/r8/nlswork, clear
(National Longitudinal Survey.  Young Women 14-26 years of age in 1968)
. generate age2=age^2
(24 missing values generated)
. generate ttl_exp2 = ttl_exp^2
. generate tenure2=tenure^2
(433 missing values generated)
. generate byte black = race==2
. xtdata ln_w grade age* ttl_exp* tenure* black not_smsa south, be clear i(id)
. regress ln_w grade age* ttl_exp* tenure* black not_smsa south
```

Source	SS	df	MS		
				Number of obs =	4697
				F(10, 4686) =	450.23
Model	415.021613	10	41.5021613	Prob > F =	0.0000
Residual	431.954995	4686	.092179896	R-squared =	0.4900
				Adj R-squared =	0.4889
Total	846.976608	4696	.180361288	Root MSE =	.30361

ln_wage	Coef.	Std. Err.	t	P>|t|	[95% Conf. Interval]	
grade	.0607602	.0020006	30.37	0.000	.0568382	.0646822
age	.0323158	.0087251	3.70	0.000	.0152105	.0494211
age2	-.0005997	.0001429	-4.20	0.000	-.0008799	-.0003194
(output omitted)						
south	-.0993378	.010136	-9.80	0.000	-.1192091	-.0794665
_cons	.3339113	.1210434	2.76	0.006	.0966093	.5712133

The output is the same as produced by xtreg, be; the reported R^2 is the R^2 between. There are no time savings in using xtdata followed by just one regress. The use of xtdata is justified when you intend to explore the specification of the model by running many alternative regressions.

◁

❑ Technical Note

It is important that when using xtdata, you eliminate any variables that you do not intend to use and that have missing values. xtdata follows a casewise-deletion rule, which means that an observation is excluded from the conversion if it is missing on any of the variables. In the example above, we specified that the variables be converted on the command line. Alternatively, we could drop the variables first, and it might even be useful to preserve our estimation sample:

```
. use http://www.stata-press.com/data/r8/nlswork, clear
(National Longitudinal Survey.  Young Women 14-26 years of age in 1968)
. generate age2 = age^2
(24 missing values generated)
. generate ttl_exp2 = ttl_exp^2
. generate tenure2 = tenure^2
(433 missing values generated)
. generate byte black = race==2
```

```
. keep id year ln_w grade age* ttl_exp* tenure* black not_smsa south

. tsset id year
        panel variable:  idcode, 1 to 5159
        time variable:   year, 68 to 88, but with gaps

. save xtdatasmpl
file xtdatasmpl.dta saved
```

❑

▷ Example

xtdata with the fe option converts the data so that results are equivalent to estimating by using xtreg with the fe option.

```
. use http://www.stata-press.com/data/r8/xtdatasmpl, clear
(National Longitudinal Survey.  Young Women 14-26 years of age in 1968)

. xtdata, fe

. regress ln_w grade age* ttl_exp* tenure* black not_smsa south
```

Source	SS	df	MS		Number of obs = 28091
					F(9, 28081) = 651.21
Model	412.443881	9	45.8270979		Prob > F = 0.0000
Residual	1976.12232	28081	.07037222		R-squared = 0.1727
					Adj R-squared = 0.1724
Total	2388.5662	28090	.085032617		Root MSE = .26528

ln_wage	Coef.	Std. Err.	t	P>|t|	[95% Conf. Interval]
grade	-.0147051	4.97e+08	-0.00	1.000	-9.75e+08 9.75e+08
age	.0359987	.0030904	11.65	0.000	.0299414 .0420559
age2	-.000723	.0000486	-14.88	0.000	-.0008183 -.0006277
(output omitted)					
south	-.0606309	.0099763	-6.08	0.000	-.0801849 -.0410769
_cons	1.221668	6.23e+09	0.00	1.000	-1.22e+10 1.22e+10

The coefficients reported by regress after xtdata, fe are the same as those reported by xtreg, fe, but the standard errors are slightly smaller. This is because no adjustment has been made to the estimated covariance matrix for the estimation of the person means. The difference is small, however, and results are adequate for a specification search.

◁

▷ Example

To use xtdata, re, you must specify the ratio $\sigma_\nu/\sigma_\epsilon$, the ratio of the standard deviations of the random effect and pure residual. Merely to show the relationship of regress after xtdata, re to xtreg, re, we will specify this ratio as $.25790313/.29069544 = .88719358$, which is the number xtreg reports when the model is fitted from the outset; see the random-effects example in [XT] **xtreg**. For specification-search purposes, however, it is adequate to specify this number more crudely, and, in fact, in performing the specification search for this manual entry, we used ratio(1).

```
. use http://www.stata-press.com/data/r8/xtdatasmpl, clear
(National Longitudinal Survey.  Young Women 14-26 years of age in 1968)
```

```
. xtdata, clear re ratio(.88719358)
```

		theta		
min	5%	median	95%	max
0.2520	0.2520	0.5499	0.7016	0.7206

xtdata reports the distribution of θ based on the specified ratio. If these were balanced data, θ would have been constant.

When running regressions with these data, you must specify the noconstant option and include the variable constant:

```
. regress ln_w grade age* ttl_exp* tenure* black not_smsa south constant, noconstant
```

Source	SS	df	MS		Number of obs = 28091
					F(11, 28080) =14303.11
Model	13272.3241	11	1206.57492		Prob > F = 0.0000
Residual	2368.75918	28080	.084357521		R-squared = 0.8486
					Adj R-squared = 0.8485
Total	15641.0833	28091	.556800517		Root MSE = .29044

ln_wage	Coef.	Std. Err.	t	P>|t|	[95% Conf. Interval]
grade	.0646499	.0017811	36.30	0.000	.0611588 .068141
age	.0368059	.0031195	11.80	0.000	.0306915 .0429204
age2	-.0007133	.00005	-14.27	0.000	-.0008113 -.0006153
(output omitted)					
south	-.0868927	.0073031	-11.90	0.000	-.1012072 -.0725781
constant	.238721	.0494688	4.83	0.000	.1417598 .3356822

Results are the same coefficients and standard errors that xtreg, re previously estimated. The summaries at the top, however, should be ignored. These summaries are expressed in terms of equation (4) of [XT] **xtreg**, and, moreover, for a model without a constant.

◁

❏ Technical Note

Obviously, some caution is required in using xtdata. The following guidelines will help:

1. xtdata is intended for use during the specification search phase of analysis only. Final results should be estimated with xtreg on unconverted data.

2. After converting the data, you may use regress to obtain estimates of the coefficients and their standard errors. In the case of regress after xtdata, fe, the standard errors are too small, but only slightly.

3. You may loosely interpret the coefficient's significance tests and confidence intervals. However, for results after xtdata, fe and re, a wrong (but very close to correct) distribution is being assumed.

4. You should ignore the summary statistics reported at the top of regress's output.

5. After converting the data, you may form linear, but not nonlinear, combinations of regressors; that is, if your data contained age, it would not be correct to convert the data and then form age squared. All nonlinear transformations should be done before conversion. (For xtdata, be, you can get away with forming nonlinear combinations ex post, but the results will not be exact.)

❏

❏ Technical Note

The xtdata command can be used to assist in examining data, especially with scatter.

```
. use http://www.stata-press.com/data/r8/xtdatasmpl, clear
(National Longitudinal Survey.  Young Women 14-26 years of age in 1968)
. xtdata, be
. scatter ln_wage age, title(Between data) msymbol(o) msize(tiny)
```

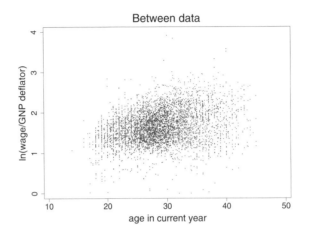

```
. use http://www.stata-press.com/data/r8/xtdatasmpl, clear
(National Longitudinal Survey.  Young Women 14-26 years of age in 1968)
. xtdata, fe
. scatter ln_wage age, title(Within data) msymbol(o) msize(tiny)
```

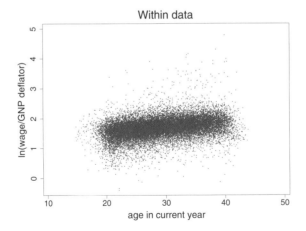

(Continued on next page)

```
. use http://www.stata-press.com/data/r8/xtdatasmpl, clear
(National Longitudinal Survey.  Young Women 14-26 years of age in 1968)
. scatter ln_wage age, title(Overall data) msymbol(o) msize(tiny)
```

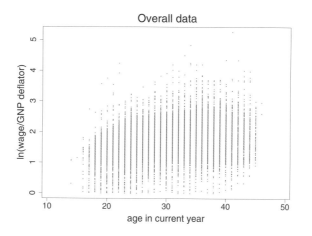

Methods and Formulas

xtdata is implemented as an ado-file.

(This is a continuation of the *Methods and Formulas* of [XT] **xtreg**.)

xtdata, be, fe, and re transform the data according to (2), (3), and (4), respectively, of [XT] **xtreg**, except that xtdata, fe adds back in the overall mean, thus forming the transformation

$$\mathbf{x}_{it} - \overline{x}_i + \overline{\overline{x}}$$

xtdata, re requires the user to specify r as an estimate of $\sigma_\nu/\sigma_\epsilon$. θ_i is calculated from

$$\theta_i = 1 - \frac{1}{\sqrt{T_i r^2 + 1}}$$

Also See

Complementary: [XT] **xtreg**, [XT] **xtregar**,
                      [R] **regress**

Background: [XT] **xt**

Title

> **xtdes** — Describe pattern of xt data

Syntax

xtdes [if *exp*] [in *range*] [, $\underline{p}$atterns(#) i(*varname$_i$*) t(*varname$_t$*)]

by ... : may be used with xtdes; see [R] **by**.

Description

xtdes describes the participation pattern of cross-sectional time-series (xt) data.

Options

patterns(#) specifies the maximum number of participation patterns to be reported; patterns(9) is the default. Specifying patterns(50) would list up to 50 patterns. Specifying patterns(1000) is taken to mean patterns(∞); all the patterns will be listed.

i(*varname$_i$*) specifies the variable name corresponding to i in x_{it}. You can specify the i() option the first time you estimate, or you can use the iis command to set i() beforehand. Note that it is not necessary to specify i() if the data have been previously tsset, or if iis has been previously specified—in these cases, the group variable is taken from the previous setting. See [XT] **xt**.

t(*varname$_t$*) specifies the variable name corresponding to t in x_{it}. You can specify the t() option the first time you estimate, or you can use the tis command to set t() beforehand. Note that it is not necessary to specify t() if the data have been previously tsset, or if tis has been previously specified—in these cases, the group variable is taken from the previous setting. See [XT] **xt**.

Remarks

If you have not read [XT] **xt**, please do so.

xtdes does not have a simple-data counterpart. It describes the cross-sectional and time-series aspects of the data in memory.

▷ Example

In [XT] **xt**, we introduced data based on a subsample of the NLSY data on young women aged 14–26 in 1968. Here is a description of the data used in many of the [XT] **xt** examples:

(*Continued on next page*)

48

```
. use http://www.stata-press.com/data/r8/nlswork
(National Longitudinal Survey.  Young Women 14-26 years of age in 1968)

. xtdes, i(id) t(year)
   idcode:  1, 2, ..., 5159                              n =      4711
     year:  68, 69, ..., 88                              T =        15
            Delta(year) = 1; (88-68)+1 = 21
            (idcode*year uniquely identifies each observation)
Distribution of T_i:    min     5%    25%      50%      75%    95%    max
                          1      1      3        5        9     13     15

      Freq.   Percent   Cum.  | Pattern
      ────────────────────────┼─────────────────────
        136      2.89   2.89  | 1....................
        114      2.42   5.31  | ....................1
         89      1.89   7.20  | ................1.11
         87      1.85   9.04  | ..................11
         86      1.83  10.87  | 111111.1.11.1.11.1.11
         61      1.29  12.16  | ............11.1.11
         56      1.19  13.35  | 11..................
         54      1.15  14.50  | ......1.11.1.11.1.11
         54      1.15  15.64  | .............1.1.11
       3974     84.36 100.00  | (other patterns)
      ────────────────────────┼─────────────────────
       4711    100.00         | XXXXXX.X.XX.X.XX.X.XX
```

xtdes tells us that we have 4,711 women in our data and that the idcode that identifies each ranges from 1 to 5,159. We are also told that the maximum number of individual years over which we observe any woman is 15. (The year variable, however, ranges over 21 years.) We are reassured that idcode and year, taken together, uniquely identify each observation in our data. We are also shown the distribution of T_i; 50% of our women are observed 5 years or less. Only 5% of our women are observed for 13 years or more.

Finally, we are shown the participation pattern. A 1 in the pattern means one observation that year; a dot means no observation. The largest fraction of our women, but still only 2.89%, was observed in the single year 1968 and not thereafter; the next largest fraction was observed in 1988 but not before; and the next largest fraction was observed in 1985, 1987, and 1988.

At the bottom is the sum of the participation patterns, including the patterns that were not shown. We can see that none of the women were observed in six of the years (there are six dots). (The survey was not administered in those six years.)

If we wished, we could see more of the patterns by specifying the patterns() option, and, in fact, we could see all the patterns by specifying patterns(1000).

◁

▷ Example

The strange participation patterns shown above have to do with our subsampling of the data, and not with the administrators of the survey. As an aside, here are the data from which we drew the sample used in the [XT] **xt** examples:

(Continued on next page)

```
. xtdes
  idcode:  1, 2, ..., 5159                                    n =        5159
    year:  68, 69, ..., 88                                    T =          15
           Delta(year) = 1; (88-68)+1 = 21
           (idcode*year does not uniquely identify observations)
  Distribution of T_i:    min      5%     25%      50%     75%    95%     max
                            1       2      11       15      16     19      30

     Freq.   Percent   Cum.  │  Pattern
   ─────────────────────────────────────────────────────
     1034     20.04   20.04  │  111111.1.11.1.11.1.11
      153      2.97   23.01  │  1....................
      147      2.85   25.86  │  112111.1.11.1.11.1.11
      130      2.52   28.38  │  111112.1.11.1.11.1.11
      122      2.36   30.74  │  111211.1.11.1.11.1.11
      113      2.19   32.93  │  11...................
       84      1.63   34.56  │  111111.1.11.1.11.1.12
       79      1.53   36.09  │  111111.1.12.1.11.1.11
       67      1.30   37.39  │  111111.1.11.1.11.1.1.
     3230     62.61  100.00  │  (other patterns)
   ─────────────────────────────────────────────────────
     5159    100.00          │  XXXXXX.X.XX.X.XX.X.XX
```

Note that we have multiple observations per year. In the pattern, 2 is used to indicate that a woman appears twice in the year, 3 to indicate 3 times, and so on—X is used to mean 10 or more should that be necessary.

In fact, this is a dataset that was itself extracted from the NLSY, in which t is not time but job number. In order to simplify exposition, we made a simpler dataset by selecting the last job in each year.

◁

Methods and Formulas

xtdes is implemented as an ado-file.

Also See

Related: [XT] **xtsum**, [XT] **xttab**

Background: [XT] **xt**

Title

> **xtfrontier** — Stochastic frontier models for panel data

Syntax

Time-invariant model

> xtfrontier *depvar* $\left[\textit{varlist}\right]$ $\left[\textit{weight}\right]$ $\left[\texttt{if } \textit{exp}\right]$ $\left[\texttt{in } \textit{range}\right]$, $\left[\texttt{ti } \texttt{i}(\textit{varname})\right.$
>
> cost <u>no</u>constant <u>constr</u>aints(*numlist*) <u>l</u>evel(*#*) <u>nodif</u>ficult <u>nolog</u>
>
> *maximize_options* $\left.\right]$

Time-varying decay model

> xtfrontier *depvar* $\left[\textit{varlist}\right]$ $\left[\textit{weight}\right]$ $\left[\texttt{if } \textit{exp}\right]$ $\left[\texttt{in } \textit{range}\right]$, $\left[\texttt{tvd } \texttt{i}(\textit{varname})\right.$
>
> t(*varname*) cost <u>no</u>constant <u>constr</u>aints(*numlist*) <u>l</u>evel(*#*)
>
> <u>nodif</u>ficult <u>nolog</u> *maximize_options* $\left.\right]$

by ... : may be used with xtfrontier; see [R] **by**.

You must tsset your data in order to use time-series operators; see [TS] **tsset**.

depvar and all *varlist*s may contain time-series operators; see [U] **14.4.3 Time-series varlists**.

fweights and iweights are allowed; see [U] **14.1.6 weight**.

xtfrontier shares the features of all estimation commands; see [U] **23 Estimation and post-estimation commands**.

Syntax for predict

> predict $\left[\textit{type}\right]$ *newvarname* $\left[\texttt{if } \textit{exp}\right]$ $\left[\texttt{in } \textit{range}\right]$ $\left[\texttt{, } \textit{statistic}\right]$

where *statistic* is

xb	$\mathbf{x}_j\mathbf{b}$, fitted value (the default)
stdp	standard error of the prediction
u	estimates of minus the natural log of the technical efficiency via $E\left(u_{it} \mid \epsilon_{it}\right)$
m	estimates of minus the natural log of the technical efficiency via $M\left(u_{it} \mid \epsilon_{it}\right)$
te	estimates of the technical efficiency via $E\left\{\exp(-su_{it}) \mid \epsilon_{it}\right\}$

where

$$s = \begin{cases} 1, & \text{for production functions} \\ -1, & \text{for cost functions} \end{cases}$$

Description

xtfrontier fits stochastic production or cost frontier models for panel data. More precisely, xtfrontier contains estimators for the parameters of a linear model with a disturbance generated by specific mixture distributions. These estimators have been used almost exclusively to estimate production and cost frontiers, and it is in this context that the estimators provided by xtfrontier are presented.

The disturbance term in a stochastic frontier model is assumed to have two components. One component is assumed to have a strictly non-negative distribution, and the other component is assumed to have a symmetric distribution. In the econometrics literature, the non-negative component is often referred to as the inefficiency term, and the component with the symmetric distribution is often referred to as the idiosyncratic error. We use this terminology in the discussion that follows. xtfrontier permits two different parameterizations of the inefficiency term: a time-invariant model and the Battese–Coelli (1992) parameterization of time effects. In the time-invariant model, the inefficiency term is assumed to have a truncated-normal distribution. In the Battese–Coelli (1992) parameterization of time effects, the inefficiency term is modeled as a truncated-normal random variable multiplied by a specific function of time. In both models, the idiosyncratic error term is assumed to have a normal distribution. The only panel-specific effect is the random inefficiency term.

See Kumbhakar and Lovell (2000) for a detailed introduction to frontier analysis.

Options

ti specifies that the parameters of the time-invariant technical inefficiency model are to be estimated.

tvd specifies that the parameters of the time-varying decay model are to be estimated.

i(*varname*) specifies the variable name that contains the unit to which the observation belongs. You can specify the i() option the first time you estimate, or you can use the iis command to set i() beforehand. Note that it is not necessary to specify i() if the data have been previously tsset, or if iis has been previously specified—in these cases, the group variable is taken from the previous setting. See [XT] **xt**.

t(*varname*) specifies the variable name that contains the time period to which the observation belongs. You can specify the t() option the first time you estimate, or you can use the tis command to set t() beforehand. Note that it is not necessary to specify t() if the data have been previously tsset, or if tis has been previously specified—in these cases, the group variable is taken from the previous setting. See [XT] **xt**.

cost specifies that the frontier model will be fitted in terms of a cost function instead of a production function. By default, xtfrontier fits a production frontier model.

noconstant suppresses the constant term (intercept) from the model.

constraints(*numlist*) specifies the constraint numbers of the linear constraints to be applied during estimation. The default is to perform unconstrained estimation. Constraints are specified using the constraint command; see [R] **constraint**.

level(*#*) specifies the confidence level, in percent, for confidence intervals. The default is level(95) or as set by set level; see [U] **23.6 Specifying the width of confidence intervals**.

nodifficult specifies that the maximization option difficult is not to be utilized. By default, difficult is specified since these models often have problems converging. If your model is taking many iterations to converge, try specifying nodifficult because sometimes changing this specification will help.

nolog suppresses the iteration log.

maximize_options control the maximization process; see [R] **maximize**. You will seldom need to specify any of the maximize options, except for `iterate(0)` and possibly `from()`. If the iteration log shows many "not concave" messages and it is taking many iterations to converge, try specifying better starting values by using the `from()` option.

Options for predict

`xb`, the default, calculates the linear prediction.

`stdp` requests prediction of the standard error of linear prediction.

`u` produces estimates of minus the natural log of the technical efficiency via $E\left(u_{it} \mid \epsilon_{it}\right)$.

`m` produces estimates of minus the natural log of the technical efficiency via the mode, $M\left(u_{it} \mid \epsilon_{it}\right)$.

`te` produces estimates of the technical efficiency via $E\left\{\exp(-su_{it}) \mid \epsilon_{it}\right\}$.

Remarks

Remarks are presented under the headings

Introduction
Time-invariant model
Time-varying decay model

Introduction

Stochastic production frontier models were introduced by Aigner, Lovell, and Schmidt (1977) and Meeusen and van den Broeck (1977). Since then, stochastic frontier models have become a popular subfield in econometrics; see Kumbhakar and Lovell (2000) for an introduction. `xtfrontier` fits two stochastic frontier models with distinct specifications of the inefficiency term. `xtfrontier` can fit both production- and cost-frontier models.

Let's review the nature of the stochastic frontier problem. Suppose that a producer has a production function $f(\mathbf{z}_{it}, \beta)$. In a world without error or inefficiency, in time t, the ith firm would produce

$$q_{it} = f(\mathbf{z}_{it}, \beta)$$

A fundamental element of stochastic frontier analysis is that each firm potentially produces less than it might due to a degree of inefficiency. Specifically,

$$q_{it} = f(\mathbf{z}_{it}, \beta)\xi_{it}$$

where ξ_{it} is the level of efficiency for firm i at time t; ξ_i must be in the interval $(0, 1]$. If $\xi_{it} = 1$, then the firm is achieving the optimal output with the technology embodied in the production function $f(\mathbf{z}_{it}, \beta)$. When $\xi_{it} < 1$, the firm is not making the most of the inputs $\mathbf{z}_{it}$ given the technology embodied in the production function $f(\mathbf{z}_{it}, \beta)$. Since the output is assumed to be strictly positive (i.e., $q_{it} > 0$), the degree of technical efficiency is assumed to be strictly positive (i.e., $\xi_{it} > 0$).

Output is also assumed to be subject to random shocks, implying that

$$q_{it} = f(\mathbf{z}_{it}, \beta)\xi_{it}\exp(v_{it})$$

Taking the natural log of both sides yields

$$\ln(q_{it}) = \ln\{f(\mathbf{z}_{it}, \beta)\} + \ln(\xi_{it}) + v_{it}$$

Assuming that there are k inputs and that the production function is linear in logs, defining $u_{it} = -\ln(\xi_{it})$ yields

$$\ln(q_{it}) = \beta_0 + \sum_{j=1}^{k} \beta_j \ln(z_{jit}) + v_{it} - u_{it} \tag{1}$$

Since u_{it} is subtracted from $\ln(q_{it})$, restricting $u_{it} \geq 0$ implies that $0 < \xi_{it} \leq 1$ as specified above.

Kumbhakar and Lovell (2000) provide a detailed version of the above derivation, and they show that performing an analogous derivation in the dual-cost function problem allows one to specify the problem as

$$\ln(c_{it}) = \beta_0 + \beta_q \ln(q_{it}) + \sum_{j=1}^{k} \beta_j \ln(p_{jit}) + v_{it} - s u_{it} \tag{2}$$

where q_{it} is output, the z_{jit} are input quantities , c_{it} is cost, and the p_{jit} are input prices, and

$$s = \begin{cases} 1, & \text{for production functions} \\ -1, & \text{for cost functions} \end{cases}$$

Intuitively, the inefficiency effect is required to lower output or raise expenditure, depending on the specification.

❑ Technical Note

The model that xtfrontier actually fits is of the form

$$y_{it} = \beta_0 + \sum_{j=1}^{k} \beta_j x_{jit} + v_{it} - s u_{it}$$

so in the context of the discussion above, $y_{it} = \ln(q_{it})$ and $x_{jit} = \ln(z_{jit})$ for a production function, and for a cost function, $y_{it} = \ln(c_{it})$, the x_{jit} are the $\ln(p_{jit})$, and $\ln(q_{it})$. It is incumbent upon the user to perform the natural logarithm transformation of the data prior to estimation if the estimation results are to be correctly interpreted in the context of a stochastic frontier production or cost model. xtfrontier does not perform any transformations on the data.

❑

Equation (2) is a variant of a panel-data model in which v_{it} is the idiosyncratic error and u_{it} is a time-varying panel-level effect. Much of the literature on this model has focused on deriving estimators for different specifications of the u_{it} term. Kumbhakar and Lovell (2000) provide a survey of this literature.

xtfrontier provides estimators for two different specifications of u_{it}. To facilitate the discussion, let $N^+(\mu, \sigma^2)$ denote the truncated-normal distribution, which is truncated at zero with mean μ and variance σ^2, and let $\overset{\text{iid}}{\sim}$ stand for independently and identically distributed.

Consider the simplest specification in which u_{it} is a time-invariant truncated-normal random variable. In the time-invariant model, $u_{it} = u_i$, $u_i \overset{\text{iid}}{\sim} N^+(\mu, \sigma_\mu^2)$, $v_{it} \overset{\text{iid}}{\sim} N(0, \sigma_v^2)$, and u_i and v_{it} are distributed independently of each other and the covariates in the model. Specifying the option $\texttt{ti}$ causes $\texttt{xtfrontier}$ to estimate the parameters of this model.

In the time-varying decay specification,

$$u_{it} = \exp\{-\eta(t - T_i)\}u_i$$

where T_i is the last time period in the ith panel, η is the decay parameter, $u_i \overset{\text{iid}}{\sim} N^+(\mu, \sigma_\mu^2)$, $v_{it} \overset{\text{iid}}{\sim} N(0, \sigma_v^2)$, and u_i and v_{it} are distributed independently of each other and the covariates in the model. Specifying the option $\texttt{tvd}$ causes $\texttt{xtfrontier}$ to estimate the parameters of this model.

Time-invariant model

▷ Example

To illustrate the $\texttt{xtfrontier}$ command, we begin with an example using fictional data. In this example, fictional firms produce a fictional product called a widget, using a constant-returns-to-scale technology. We have 948 observations; 91 firms, with 6–14 observations per firm. Our dataset contains variables representing the quantity of widgets produced, the number of machine hours used in production, the number of labor hours used in production, and three additional variables that are the natural logarithm transformations of the three aforementioned variables.

We fit a time-invariant model using the transformed variables:

(Continued on next page)

```
. use http://www.stata-press.com/data/r8/xtfrontier1

. xtfrontier lnwidgets lnmachines lnworkers, ti

Iteration 0:   log likelihood = -1473.8703
Iteration 1:   log likelihood = -1473.0565
Iteration 2:   log likelihood = -1472.6155
Iteration 3:   log likelihood =  -1472.607
Iteration 4:   log likelihood = -1472.6069
```

Time-invariant inefficiency model				Number of obs	=	948
Group variable (i): id				Number of groups	=	91
Time variable (t): t				Obs per group: min =		6
				avg =		10.4
				max =		14
				Wald chi2(2)	=	661.76
Log likelihood = -1472.6069				Prob > chi2	=	0.0000

lnwidgets	Coef.	Std. Err.	z	P>|z|	[95% Conf. Interval]	
lnmachines	.2904551	.0164219	17.69	0.000	.2582688	.3226415
lnworkers	.2943333	.0154352	19.07	0.000	.2640808	.3245858
_cons	3.030983	.1441022	21.03	0.000	2.748548	3.313418
/mu	1.125667	.6479217	1.74	0.082	-.144236	2.39557
/lnsigma2	1.421979	.2672745	5.32	0.000	.898131	1.945828
/ilgtgamma	1.138685	.3562642	3.20	0.001	.4404204	1.83695
sigma2	4.145318	1.107938			2.455011	6.999424
gamma	.7574382	.0654548			.6083592	.8625876
sigma_u2	3.139822	1.107235			.9696821	5.309962
sigma_v2	1.005496	.0484143			.9106055	1.100386

In addition to the coefficients, the output reports estimates for the parameters `sigma_v2`, `sigma_u2`, `gamma`, `sigma2`, `ilgtgamma`, `lnsigma2`, and `mu`. `sigma_v2` is the estimate of σ_v^2. `sigma_u2` is the estimate of σ_u^2. `gamma` is the estimate of $\gamma = \sigma_u^2/\sigma_S^2$. `sigma2` is the estimate of $\sigma_S^2 = \sigma_v^2 + \sigma_u^2$. Since γ must be between 0 and 1, the optimization is parameterized in terms of the inverse logit of γ, and this estimate is reported as `ilgtgamma`. Since σ_S^2 must be positive, the optimization is parameterized in terms of $\ln(\sigma_S^2)$, whose estimate is reported as `lnsigma2`. Finally, `mu` is the estimate of μ.

◁

❑ Technical Note

Our simulation results indicate that this estimator requires relatively large samples in order to achieve any reasonable degree of precision in the estimates of μ and $\sigma_\mu{}^2$.

❑

Time-varying decay model

`xtfrontier, tvd` provides maximum likelihood estimates for the parameters of the time-varying decay model. In this model, the inefficiency effects are modeled as

$$u_{it} = \exp\bigl\{-\eta(t - T_i)\bigr\} u_i$$

where $u_i \stackrel{\text{iid}}{\sim} N^+(\mu, \sigma_\mu^2)$.

When $\eta > 0$, the degree of inefficiency is decreasing over time; when $\eta < 0$, the degree of inefficiency is increasing over time. Since $t = T_i$ in the last period, the last period for firm i contains the base level of inefficiency for that firm. If $\eta > 0$, then the level of inefficiency decays toward the base level. If $\eta < 0$, then the level of inefficiency increases to the base level.

▷ Example

When $\eta = 0$, the time-varying decay model reduces to the time-invariant model. The following example illustrates this property. We will also demonstrate how to specify constraints and starting values in these models.

Let's begin by fitting the time-varying decay model on the same data that were used in the previous example for the time-invariant model.

```
. xtfrontier lnwidgets lnmachines lnworkers, tvd
Iteration 0:   log likelihood = -1551.3798  (not concave)
Iteration 1:   log likelihood = -1502.2637
Iteration 2:   log likelihood = -1476.3093  (not concave)
Iteration 3:   log likelihood = -1472.9845
Iteration 4:   log likelihood = -1472.5365
Iteration 5:   log likelihood =  -1472.529
Iteration 6:   log likelihood = -1472.5289
```

```
Time-varying decay inefficiency model           Number of obs      =       948
Group variable (i): id                          Number of groups   =        91
Time variable (t): t                            Obs per group: min =         6
                                                               avg =      10.4
                                                               max =        14
                                                Wald chi2(2)       =    661.93
Log likelihood  = -1472.5289                    Prob > chi2        =    0.0000
```

lnwidgets	Coef.	Std. Err.	z	P>|z|	[95% Conf. Interval]	
lnmachines	.2907555	.0164376	17.69	0.000	.2585384	.3229725
lnworkers	.2942412	.0154373	19.06	0.000	.2639846	.3244978
_cons	3.028939	.1436046	21.09	0.000	2.74748	3.310399
/mu	1.110831	.6452809	1.72	0.085	-.1538967	2.375558
/eta	.0016764	.00425	0.39	0.693	-.0066535	.0100064
/lnsigma2	1.410723	.2679485	5.26	0.000	.885554	1.935893
/ilgtgamma	1.123982	.3584243	3.14	0.002	.4214828	1.82648
sigma2	4.098919	1.098299			2.424327	6.930228
gamma	.7547265	.0663495			.603838	.8613419
sigma_u2	3.093563	1.097606			.9422943	5.244832
sigma_v2	1.005356	.0484079			.9104785	1.100234

The estimate of η is very close to zero, and the other estimates are not too far from those of the time-invariant model.

We can use constraint define to constrain $\eta = 0$ and obtain the same results produced by the time-invariant model. Although there is only one statistical equation to be estimated in this model, the model fits five of Stata's [R] **ml** equations; see [R] **ml** or Gould and Sribney (1999). The equation names can be seen by listing the matrix of estimated coefficients.

```
. matrix list e(b)

e(b)[1,7]
        lnwidgets:  lnwidgets:  lnwidgets:   lnsigma2:  ilgtgamma:        mu:
        lnmachines   lnworkers       _cons       _cons       _cons      _cons
y1     .29075546    .2942412   3.0289395   1.4107233   1.1239816  1.1108307

             eta:
            _cons
y1      .00167642
```

To constrain a parameter to a particular value in any equation, except the first equation, it is necessary to specify both the equation name and the parameter name using the syntax

constraint define # [*eqname*]_b[*varname*] = *value*   or

constraint define # [*eqname*]*coefficient* = *value*

where *eqname* gives the equation name and *varname* gives the name of variable in a linear equation, and *coefficient* refers to any parameter that has been estimated. More elaborate specifications with expressions are possible; see the example with constant returns to scale below, and see [R] **constraint** for general reference.

Suppose that we impose the constraint $\eta = 0$; we get the same results as those reported above for the time-invariant model, except for some minute differences attributable to an alternate convergence path in the optimization.

```
. constraint define 1 [eta]_cons = 0

. xtfrontier lnwidgets lnmachines lnworkers, tvd constraints(1)

Iteration 0:   log likelihood = -1540.7124   (not concave)
Iteration 1:   log likelihood = -1515.7726
Iteration 2:   log likelihood = -1473.0162
Iteration 3:   log likelihood = -1472.9223
Iteration 4:   log likelihood = -1472.6254
Iteration 5:   log likelihood =  -1472.607
Iteration 6:   log likelihood = -1472.6069
```

Time-varying decay inefficiency model	Number of obs	=	948
Group variable (i): id	Number of groups	=	91
Time variable (t): t	Obs per group: min =		6
	avg =		10.4
	max =		14
	Wald chi2(2)	=	661.76
Log likelihood = -1472.6069	Prob > chi2	=	0.0000

lnwidgets	Coef.	Std. Err.	z	P>|z|	[95% Conf. Interval]	
lnmachines	.2904551	.0164219	17.69	0.000	.2582688	.3226414
lnworkers	.2943332	.0154352	19.07	0.000	.2640807	.3245857
_cons	3.030963	.1440995	21.03	0.000	2.748534	3.313393
/mu	1.125507	.6480444	1.74	0.082	-.1446369	2.39565
/eta	0	.	.	.	.	.
/lnsigma2	1.422039	.2673128	5.32	0.000	.8981155	1.945962
/ilgtgamma	1.138764	.3563076	3.20	0.001	.4404135	1.837114
sigma2	4.145565	1.108162			2.454972	7.000366
gamma	.7574526	.0654602			.6083575	.862607
sigma_u2	3.140068	1.107459			.9694878	5.310649
sigma_v2	1.005496	.0484143			.9106057	1.100386

◁

Saved Results

xtfrontier saves in e():

Scalars

e(N)	number of observations	e(chi2)	χ^2
e(k)	number of estimated parameters	e(sigma_u)	standard deviation of
e(k_eq)	number of equations		technical inefficiency
e(k_dv)	number of dependent variables	e(sigma_v)	standard deviation of random error
e(df_m)	model degrees of freedom	e(rank)	rank of VCE
e(ll)	log likelihood	e(p)	model significance
e(rc)	return code	e(ic)	number of iterations

Macros

e(cmd)	xtfrontier	e(wtype)	weight type
e(depvar)	name of dependent variable	e(wexp)	weight expression
e(title)	name of model	e(chi2type)	Wald; type of model χ^2 test
e(function)	production or cost	e(opt)	ml
e(model)	ti, after time-invariant model;	e(user)	program used to implement ml
	tvd, after time-varying decay model	e(crittype)	optimization criterion
e(ivar)	variable denoting groups	e(predict)	program used to implement
e(tvar)	variable denoting time periods		predict

Matrices

e(b)	coefficient vector	e(V)	variance–covariance matrix
e(ilog)	iteration log (up to 20 iterations)		of the estimators

Functions

e(sample)	marks estimation sample

Methods and Formulas

xtfrontier is implemented as an ado-file.

xtfrontier fits stochastic frontier models for panel data that can be expressed as

$$y_{it} = \beta_0 + \sum_{j=1}^{k} \beta_j x_{jit} + v_{it} - s u_{it}$$

where y_{it} is the natural logarithm of output and the x_{jit} are the natural logarithm of the input quantities for the production efficiency problem, y_{it} is the natural logarithm of costs and the x_{it} are the natural logarithm of input prices for the cost efficiency problem, and

$$s = \begin{cases} 1, & \text{for production functions} \\ -1, & \text{for cost functions} \end{cases}$$

For the time-varying decay model, the log-likelihood function is derived as

$$\ln L = -\frac{1}{2}\left(\sum_{i=1}^{N}T_i\right)\{\ln(2\pi)+\ln(\sigma_S^2)\} - \frac{1}{2}\sum_{i=1}^{N}(T_i-1)\ln(1-\gamma)$$

$$-\frac{1}{2}\sum_{i=1}^{N}\ln\left\{1+\left(\sum_{t=1}^{T_i}\eta_{it}^2-1\right)\gamma\right\} - N\ln\{1-\Phi(-\widetilde{z})\} - \frac{1}{2}N\widetilde{z}^2$$

$$+\sum_{i=1}^{N}\ln\{1-\Phi(-z_i^*)\} + \frac{1}{2}\sum_{i=1}^{N}z_i^{*2} - \frac{1}{2}\sum_{i=1}^{N}\sum_{t=1}^{T_i}\frac{\epsilon_{it}^2}{(1-\gamma)\sigma_S^2}$$

where $\sigma_S = (\sigma_u^2+\sigma_v^2)^{1/2}$, $\gamma = \sigma_u^2/\sigma_S^2$, $\epsilon_{it} = y_{it}-\mathbf{x}_{it}\boldsymbol{\beta}$, $\eta_{it} = \exp\{-\eta(t-T_i)\}$, $\widetilde{z} = \mu/\left(\gamma\sigma_S^2\right)^{1/2}$, $\Phi()$ is the cumulative distribution function of the standard normal distribution, and

$$z_i^* = \frac{\mu(1-\gamma)-s\gamma\sum_{t=1}^{T_i}\eta_{it}\epsilon_{it}}{\left[\gamma(1-\gamma)\sigma_S^2\left\{1+\left(\sum_{t=1}^{T_i}\eta_{it}^2-1\right)\gamma\right\}\right]^{1/2}}$$

Maximizing the above log likelihood provides estimates of the coefficients, η, μ, σ_v, and σ_u. Given these estimates, estimates for u_{it} can be obtained from the mean or the mode of the conditional distribution $f(u|\epsilon)$.

$$E(u_{it} \mid \epsilon_{it}) = \widetilde{\mu}_i + \widetilde{\sigma}_i\left\{\frac{\phi(-\widetilde{\mu}_i/\widetilde{\sigma}_i)}{1-\Phi(-\widetilde{\mu}_i/\widetilde{\sigma}_i)}\right\}$$

$$M(u_{it} \mid \epsilon_{it}) = \begin{cases} -\widetilde{\mu}_i, & \text{if } \widetilde{\mu}_i >= 0 \\ 0, & \text{otherwise} \end{cases}$$

where

$$\widetilde{\mu}_i = \frac{\mu\sigma_v^2 - s\sum_{t=1}^{T_i}\eta_{it}\epsilon_{it}\sigma_u^2}{\sigma_v^2+\sum_{t=1}^{T_i}\eta_{it}^2\sigma_u^2}$$

$$\widetilde{\sigma}_i^2 = \frac{\sigma_v^2\sigma_u^2}{\sigma_v^2+\sum_{t=1}^{T_i}\eta_{it}^2\sigma_u^2}$$

These estimates can be obtained from `predict` *newvarname*, `u` and `predict` *newvarname*, `m`, respectively, and are calculated by plugging in the estimated parameters.

`predict` *newvarname*, `te` produces estimates of the technical efficiency term. These estimates are obtained from

$$E\{\exp(-u_{it}) \mid \epsilon_{it}\} = \left[\frac{1-\Phi\{s\eta_{it}\widetilde{\sigma}_i - (\widetilde{\mu}_i/\widetilde{\sigma}_i)\}}{1-\Phi(-\widetilde{\mu}_i/\widetilde{\sigma}_i)}\right]\exp\left(-s\eta_{it}\widetilde{\mu}_i + \frac{1}{2}\eta_{it}^2\widetilde{\sigma}_i^2\right)$$

Replacing $\eta_{it} = 1$ and $\eta = 0$ in the above formulas produces the formulas for the time-invariant models.

References

Aigner, D. J., C. A. K. Lovell, and P. Schmidt. 1977. Formulation and estimation of stochastic frontier production function models. *Journal of Econometrics* 6: 21–37.

Battese, G. E. and T. J. Coelli. 1992. Frontier production functions, technical efficiency and panel data: with applications to paddy farmers in India. *Journal of Productivity Analysis* 3: 153–169.

——. 1995. A model for technical inefficiency effects in a stochastic frontier production for panel data. *Empirical Economics* 20: 325–332.

Caudill, S. B., J. M. Ford, and D. M. Gropper. 1995. Frontier estimation and firm-specific inefficiency measures in the presence of heteroskedasticity. *Journal of Business and Economic Statistics* 13(1): 105–111.

Coelli, T. J. 1995. Estimators and hypothesis tests for a stochastic frontier function: A Monte Carlo analysis. *Journal of Productivity Analysis* 6(4): 247–268.

Coelli, T. J., D. S. P. Rao, and G. E. Battese. 1988. *An Introduction to Efficiency and Productivity Analysis.* Boston: Kluwer Academic Publishers.

Gould, W. W. and W. S. Sribney. 1999. *Maximum Likelihood Estimation with Stata.* College Station, TX: Stata Press.

Kumbhakar, S. C. and C. A. K. Lovell. 2000. *Stochastic Frontier Analysis.* Cambridge: Cambridge University Press.

Meeusen, W. and J. van den Broeck. 1977. Efficiency estimation from Cobb–Douglas production functions with composed error. *International Economic Review* 18: 435–444.

Zellner, A. and N. Revankar. 1970. Generalized production functions. *Review of Economic Studies* 37: 241–250.

Also See

Complementary:	[R] **adjust**, [R] **constraint**, [R] **lincom**, [R] **lrtest**, [R] **mfx**, [R] **nlcom**, [R] **predict**, [R] **predictnl**, [R] **test**, [R] **testnl**, [R] **vce**, [TS] **tsset**
Related:	[XT] **xthtaylor**, [XT] **xtreg**, [R] **frontier**, [R] **regress**
Background:	[U] **16.5 Accessing coefficients and standard errors**, [U] **23 Estimation and post-estimation commands**, [U] **23.14 Obtaining robust variance estimates**, [R] **maximize**

Title

> **xtgee** — Fit population-averaged panel-data models using GEE

Syntax

xtgee *depvar* $\left[\textit{varlist}\right]$ $\left[\textit{weight}\right]$ $\left[\texttt{if}\ \textit{exp}\right]$ $\left[\texttt{in}\ \textit{range}\right]$ $\left[\ ,\right.$

<u>family</u>(*family*) <u>l</u>ink(*link*) <u>c</u>orr(*correlation*) nmp rgf <u>tra</u>ce

i(*varname*) t(*varname*) force <u>r</u>obust <u>sc</u>ore(*newvar*) <u>ef</u>orm <u>leve</u>l(*#*)

<u>e</u>xposure(*varname*) <u>off</u>set(*varname*) <u>nocon</u>stant <u>s</u>cale(x2 | dev | *#* | phi)

<u>tol</u>erance(*#*) <u>iter</u>ate(*#*) <u>nolog</u> <u>nodis</u>play $\left.\right]$

xtcorr $\left[\ ,\ \underline{\texttt{c}}\texttt{ompact}\ \right]$

where *family* is one of <u>b</u>inomial $\left[\textit{#} | \textit{varname}\right]$ | <u>g</u>aussian | <u>g</u>amma | <u>ig</u>aussian |

<u>n</u>binomial $\left[\textit{#}\right]$ | <u>p</u>oisson

and *link* is one of <u>i</u>dentity | <u>c</u>loglog | log | <u>logit</u> | <u>n</u>binomial | <u>opo</u>wer $\left[\textit{#}\right]$ |

<u>pow</u>er $\left[\textit{#}\right]$ | <u>p</u>robit | <u>rec</u>iprocal

and *correlation* is one of <u>ind</u>ependent | <u>exc</u>hangeable | ar *#* | <u>s</u>tationary *#* |

<u>non</u>stationary *#* | <u>uns</u>tructured | <u>f</u>ixed *matname*

For example,

    . xtgee y x1 x2, family(gauss) link(ident) corr(exchangeable) i(id)

would estimate a random-effects linear regression—note that the corr(exchangeable) option does not in general provide random effects. It actually fits an equal-correlation population-averaged model that is equal to the random-effects model for linear regression.

by . . . : may be used with xtgee; see [R] **by**.

iweights, fweights, and pweights are allowed; see [U] **14.1.6 weight**. Note that weights must be constant within panels.

xtgee shares the features of all estimation commands; see [U] **23 Estimation and post-estimation commands**.

Syntax for predict

predict $\left[\textit{type}\right]$ *newvarname* $\left[\texttt{if}\ \textit{exp}\right]$ $\left[\texttt{in}\ \textit{range}\right]$ $\left[\ ,\ \left[\ \texttt{mu} | \underline{\texttt{r}}\texttt{ate} | \texttt{xb} | \texttt{stdp}\ \right]\right.$

<u>nooff</u>set $\left.\right]$

These statistics are available both in and out of sample; type predict . . . if e(sample) . . . if wanted only for the estimation sample.

Description

xtgee fits population-averaged panel-data models. In particular, xtgee fits general linear models and allows you to specify the within-group correlation structure for the panels.

xtcorr is for use after xtgee. It displays the estimated matrix of the within-group correlations.

See [R] **logistic** and [R] **regress** for lists of related estimation commands.

Options

Options for xtgee

family(*family*) specifies the distribution of *depvar*; family(gaussian) is the default.

link(*link*) specifies the link function; the default is the canonical link for the family() specified.

corr(*correlation*) specifies the within-group correlation structure; the default corresponds to the equal-correlation model, corr(exchangeable).

When you specify a correlation structure that requires a lag, you indicate the lag after the structure's name with or without a blank; e.g., corr(ar 1) or corr(ar1).

If you specify the fixed correlation structure, you specify the name of the matrix containing the assumed correlations following the word fixed, e.g., corr(fixed myr).

nmp specifies that the divisor $N - P$ is to be used instead of the default N, where N is the total number of observations and P is the number of coefficients estimated.

rgf specifies that the robust variance estimate is multiplied by $(N - 1)/(N - P)$, where N is the total number of observations and P is the number of coefficients estimated. This option can only be used with family(gaussian) when robust is either specified or implied by the use of pweights. Using this option implies that the robust variance estimate is not invariant to the scale of any weights used.

trace specifies that the current estimates should be printed at each iteration.

i(*varname*) specifies the variable that contains the unit to which the observation belongs. You can specify the i() option the first time you estimate, or you can use the iis command to set i() beforehand. Note that it is not necessary to specify i() if the data have been previously tsset, or if iis has been previously specified—in these cases, the group variable is taken from the previous setting. See [XT] **xt**.

t(*varname*) specifies the variable that contains the time at which the observation was made. You can specify the t() option the first time you estimate, or you can use the tis command to set t() beforehand. Note that it is not necessary to specify t() if the data have been previously tsset, or if tis has been previously specified—in these cases, the group variable is taken from the previous setting. See [XT] **xt**.

xtgee does not need to know t() for the corr(independent) and corr(exchangeable) correlation structures. Whether you specify t() makes no difference in these two cases.

force specifies that estimation is to be forced even though t() is not equally spaced. This is relevant only for correlation structures that require knowledge of t(). These correlation structures require that observations be equally spaced so that calculations based on lags correspond to a constant time change. If you specify a t() variable that indicates observations are not equally spaced, xtgee will refuse to fit the (time-dependent) model. If you also specify force, xtgee will fit the model and assume that the lags based on the data ordered by t() are appropriate.

robust specifies that the Huber/White/sandwich estimator of variance is to be used in place of the default IRLS variance estimator (see *Methods and Formulas* below). This produces valid standard errors even if the correlations within group are not as hypothesized by the specified correlation structure. It does, however, require that the model correctly specifies the mean. As such, the resulting standard errors are labeled "semi-robust" instead of "robust". Note that although there is no cluster() option, results are as if there were a cluster() option and you specified clustering on i().

score(*newvar*) creates *newvar* containing $u_{it} = \partial \ln L / \partial (\mathbf{x}_{it}\boldsymbol{\beta})$, where L is the quasi-likelihood function. Note that the scores for the independent panels can be obtained by $u_i = \sum_t u_{it}$, or, equivalently in Stata, egen ui=sum(uit), by(t), assuming that you specified score(uit).

The score vector is $\partial \ln L / \partial \boldsymbol{\beta} = \sum u_{it}\mathbf{x}_{it}$, the product of *newvar* with each covariate summed over observations. See [U] **23.15 Obtaining scores**.

eform displays the exponentiated coefficients and corresponding standard errors and confidence intervals as described in [R] **maximize**. For family(binomial) link(logit) (i.e., logistic regression), exponentiation results in odds ratios; for family(poisson) link(log) (i.e., Poisson regression), exponentiated coefficients are incidence-rate ratios.

level(*#*) specifies the confidence level, in percent, for confidence intervals. The default is level(95) or as set by set level; see [U] **23.6 Specifying the width of confidence intervals**.

exposure(*varname*) and offset(*varname*) are different ways of specifying the same thing. exposure() specifies a variable that reflects the amount of exposure over which the *depvar* events were observed for each observation; ln(*varname*) with its coefficient constrained to be 1 is entered into the log-link function. offset() specifies a variable that is to be entered directly into the log-link function with its coefficient constrained to be 1; thus, exposure is assumed to be $e^{varname}$. If you were fitting a Poisson regression model, family(poisson) link(log), for instance, you would account for exposure time by specifying offset() containing the log of exposure time.

noconstant specifies that the linear predictor has no intercept term, thus forcing it through the origin on the scale defined by the link function.

scale(x2|dev|#|phi) overrides the default scale parameter. By default, scale(1) is assumed for the discrete distributions (binomial, negative binomial, and Poisson) and scale(x2) is assumed for the continuous distributions (gamma, Gaussian, and inverse Gaussian).

scale(x2) specifies that the scale parameter be set to the Pearson chi-squared (or generalized chi-squared) statistic divided by the residual degrees of freedom, which is recommended by McCullagh and Nelder (1989) as a good general choice for continuous distributions.

scale(dev) sets the scale parameter to the deviance divided by the residual degrees of freedom. This provides an alternative to scale(x2) for continuous distributions and for over- or underdispersed discrete distributions.

scale(#) sets the scale parameter to #. For example, using scale(1) in family(gamma) models results in exponential-errors regression (if you use assume independent correlation structure). Additional use of link(log) rather than the default link(power -1) for family(gamma) essentially reproduces Stata's streg, dist(exp) nohr command (see [ST] **streg**) if all the observations are uncensored (and if you again assume independent correlation structure).

scale(phi) specifies that the variance matrix should not be rescaled at all. The default scaling that xtgee applies makes results agree with other estimators, and has been recommended by McCullagh and Nelder (1989) in the context of GLM. If you are comparing results with calculations made by other software, you may find that the other packages do not offer this feature. In such cases, specifying scale(phi) should match their results.

tolerance(#) specifies the convergence criterion for the maximum change in the estimated coefficient vector between iterations; tol(1e-6) is the default, and you should never have to specify this option.

iterate(#) specifies that the maximum number of iterations allowed in fitting the model; iter(100) is the default. You should never have to specify this option.

nolog suppresses the iteration log.

nodisplay is for programmers. It suppresses the display of the header and coefficients.

Option for xtcorr

compact specifies that only the parameters (alpha) of the estimated matrix of within-group correlations rather than the entire matrix be displayed.

Options for predict

mu, the default, and rate both calculate the predicted value of *depvar*. mu takes into account the offset() or exposure() together with the denominator if the family is binomial; rate ignores those adjustments. mu and rate are equivalent if (1) you did not specify offset() or exposure() when you fitted the xtgee model, and (2) you did not specify family(binomial #) or family(binomial *varname*), which is to say, the binomial family and a denominator.

Thus, mu and rate are the same for link(identity) family(gaussian).

mu and rate are not equivalent for link(logit) family(binomial pop). In that case, mu would predict the number of positive outcomes and rate would predict the probability of a positive outcome.

mu and rate are not equivalent for link(log) family(poisson) exposure(time). In that case, mu would predict the number of events given exposure time and rate would calculate the incidence rate—the number of events given an exposure time of 1.

xb calculates the linear prediction.

stdp calculates the standard error of the linear prediction.

nooffset is relevant only if you specified offset(*varname*), exposure(*varname*), family(binomial #), or family(binomial *varname*) when you fitted the model. It modifies the calculations made by predict so that they ignore the offset or exposure variable and ignore the binomial denominator. Thus, predict ..., mu nooffset produces the same results as predict ..., rate.

Remarks

For a thorough introduction to GEE in the estimation of GLM, see Hardin and Hilbe (2003). Further information on linear models is presented in Nelder and Wedderburn (1972). Finally, there have been a number of illuminating articles on various applications of GEE in Zeger, Liang, and Albert (1988), Zeger and Liang (1986), and Liang (1987). Pendergast et al. (1996) provide a nice survey of the current methods for analyzing clustered data in regard to binary response data. Our implementation follows that of Liang and Zeger (1986).

xtgee fits generalized linear models of y_{it} with covariates $\mathbf{x}_{it}$,

$$g\{E(y_{it})\} = \mathbf{x}_{it}\boldsymbol{\beta}, \qquad y \sim F \text{ with parameters } \theta_{it}$$

for $i = 1, \ldots, m$ and $t = 1, \ldots, n_i$, where there are n_i observations for each group identifier i. In the above, $g(\ )$ is called the link function and F the distributional family. Substituting various definitions for $g(\ )$ and F results in a surprising array of models. For instance, if y_{it} is distributed Gaussian (normal) and $g(\ )$ is the identity function, we have

$$E(y_{it}) = \mathbf{x}_{it}\boldsymbol{\beta}, \qquad y \sim N(\ )$$

yielding linear regression, random-effects regression, or other regression-related models depending on what we assume for the correlation structure.

If $g(\ )$ is the logit function and y_{it} is distributed Bernoulli (binomial), we have

$$\text{logit}\big\{E(y_{it})\big\} = \mathbf{x}_{it}\boldsymbol{\beta}, \qquad y \sim \text{Bernoulli}$$

or logistic regression. If $g(\ )$ is the natural log function and y_{it} is distributed Poisson, we have

$$\ln\big\{E(y_{it})\big\} = \mathbf{x}_{it}\boldsymbol{\beta}, \qquad y \sim \text{Poisson}$$

or Poisson regression, also known as the log-linear model. Other combinations are possible.

You specify the link function using the `link()` option, the distributional family using `family()`, and the assumed within-group correlation structure using `corr()`. The allowed link functions are

Link function	xtgee option	Min. abbreviation
cloglog	link(cloglog)	l(cl)
identity	link(identity)	l(i)
log	link(log)	l(log)
logit	link(logit)	l(logi)
negative binomial	link(nbinomial)	l(nb)
odds power	link(opower)	l(opo)
power	link(power)	l(pow)
probit	link(probit)	l(p)
reciprocal	link(reciprocal)	l(rec)

Link function `cloglog` is defined as $\ln\big\{-\ln(1-y)\big\}$.

Link function `identity` is defined as $y = y$.

Link function `log` is defined as $\ln(y)$.

Link function `logit` is defined $\ln\big\{y/(1-y)\big\}$, the natural log of the odds.

Link function `nbinomial` α is defined as $\ln\big\{y/(y+\alpha)\big\}$.

Link function `opower` k is defined as $\big[\big\{y/(1-y)\big\}^k - 1\big]/k$. If $k = 0$, then this link is the same as the logit link.

Link function `power` k is defined as y^k. If $k = 0$, then this link is the same as the log link.

Link function `probit` is defined $\Phi^{-1}(y)$, where $\Phi^{-1}(\ )$ is the inverse Gaussian cumulative.

Link function `reciprocal` is defined as $1/y$.

The allowed distributional families are

Family	xtgee option	Min. abbreviation
Bernoulli/binomial	family(binomial)	f(b)
gamma	family(gamma)	f(gam)
Gaussian (normal)	family(gaussian)	f(gau)
inverse Gaussian	family(igaussian)	f(ig)
negative binomial	family(nbinomial)	f(nb)
Poisson	family(poisson)	f(p)

family(normal) is allowed as a synonym for family(gaussian).

The binomial distribution can be specified as (1) family(binomial), (2) family(binomial #), or (3) family(binomial *varname*). In case 2, # is the value of the binomial denominator N, the number of trials. Specifying family(binomial 1) is the same as specifying family(binomial); both mean that y has the Bernoulli distribution with values 0 and 1 only. In case 3, *varname* is the variable containing the binomial denominator, thus allowing the number of trials to vary across observations.

The negative binomial distribution must be specified as family(nbinomial #), where # denotes the value of the parameter α in the negative binomial distribution. The results will be conditional on this value.

You do not have to specify both family() and link(); the default link() is the canonical link for the specified family():

Family	Default link
family(binomial)	link(logit)
family(gamma)	link(reciprocal)
family(gaussian)	link(identity)
family(igaussian)	link(power -2)
family(nbinomial)	link(log)
family(poisson)	link(log)

If you do specify both family() and link(), note that not all combinations make sense. You may choose among the following combinations:

	cloglog	identity	log	logit	nbinom	opower	power	probit	reciprocal
binomial	x	x	x	x	x	x	x	x	x
gamma		x	x				x		x
Gaussian		x	x			x	x		x
inverse Gaussian		x	x				x		x
negative binomial		x	x		x		x		x
Poisson		x	x				x		x

You specify the assumed within-group correlation structure using the corr() option. The allowed correlation structures are

Correlation structure	xtgee option	Min. abbreviation
Independent	corr(independent)	c(ind)
Exchangeable	corr(exchangeable)	c(exc)
Autoregressive	corr(ar #)	c(ar #)
Stationary	corr(stationary #)	c(sta #)
Non-stationary	corr(nonstationary #)	c(non #)
Unstructured	corr(unstructured)	c(uns)
User-specified	corr(fixed *matname*)	c(fix *matname*)

Let us explain.

Call **R** the working correlation matrix for modeling the within-group correlation, a square $\max\{n_i\} \times \max\{n_i\}$ matrix. Option corr() specifies the structure of **R**. Let $\mathbf{R}_{t,s}$ denote the t, s element.

The independent structure is defined as

$$\mathbf{R}_{t,s} = \begin{cases} 1 & \text{if } t = s \\ 0 & \text{otherwise} \end{cases}$$

The corr(exchangeable) structure (corresponding to equal-correlation models) is defined as

$$\mathbf{R}_{t,s} = \begin{cases} 1 & \text{if } t = s \\ \rho & \text{otherwise} \end{cases}$$

The corr(ar g) structure is defined as the usual correlation matrix for an AR(g) model. This is sometimes called multiplicative correlation. For example, an AR(1) model is given by

$$\mathbf{R}_{t,s} = \begin{cases} 1 & \text{if } t = s \\ \rho^{|t-s|} & \text{otherwise} \end{cases}$$

The corr(stationary g) structure is a stationary(g) model. For example, a stationary(1) model is given by

$$\mathbf{R}_{t,s} = \begin{cases} 1 & \text{if } t = s \\ \rho & \text{if } |t - s| = 1 \\ 0 & \text{otherwise} \end{cases}$$

The corr(nonstationary g) structure is a nonstationary(g) model that imposes only the constraints that the elements of the working correlation matrix along the diagonal are 1 and the elements outside of the gth band are zero,

$$\mathbf{R}_{t,s} = \begin{cases} 1 & \text{if } t = s \\ \rho_{ts} & \text{if } 0 < |t - s| \le g, \rho_{ts} = \rho_{st} \\ 0 & \text{otherwise} \end{cases}$$

The corr(unstructured) imposes only the constraint that the diagonal elements of the working correlation matrix are 1.

$$\mathbf{R}_{t,s} = \begin{cases} 1 & \text{if } t = s \\ \rho_{ts} & \text{otherwise}, \rho_{ts} = \rho_{st} \end{cases}$$

The corr(fixed *matname*) specification is taken from the user-supplied matrix, so that

$$\mathbf{R} = matname$$

In this case, the correlations are not estimated from the data. The user-supplied matrix must be a valid correlation matrix with 1s on the diagonal.

Full formulas for all the correlation structures are provided in the *Methods and Formulas* below.

❑ Technical Note

Some family(), link(), and corr() combinations result in models already fitted by Stata. These are

family()	link()	corr()	Other Stata estimation command
gaussian	identity	independent	regress
gaussian	identity	exchangeable	xtreg, re (see note 1)
gaussian	identity	exchangeable	xtreg, pa
binomial	cloglog	independent	cloglog (see note 2)
binomial	cloglog	exchangeable	xtcloglog, pa
binomial	logit	independent	logit or logistic
binomial	logit	exchangeable	xtlogit, pa
binomial	probit	independent	probit (see note 3)
binomial	probit	exchangeable	xtprobit, pa
nbinomial	nbinomial	independent	nbreg (see note 4)
poisson	log	independent	poisson
poisson	log	exchangeable	xtpoisson, pa
gamma	log	independent	streg, dist(exp) nohr (see note 5)
family	*link*	independent	glm, irls (see note 6)

Notes:

1. These methods produce the same results only in the case of balanced panels; see [XT] **xt**.

2. For cloglog estimation, `xtgee` with `corr(independent)` and `cloglog` (see [R] **cloglog**) will produce the same coefficients, but the standard errors will be only asymptotically equivalent because cloglog is not the canonical link for the binomial family.

3. For probit estimation, `xtgee` with `corr(independent)` and `probit` will produce the same coefficients, but the standard errors will be only asymptotically equivalent because probit is not the canonical link for the binomial family. If the binomial denominator is not 1, the equivalent maximum-likelihood command is `bprobit`; see [R] **probit** and [R] **glogit**.

4. Fitting a negative binomial model using `xtgee` (or using `glm`) will yield results conditional on the specified value of α. The `nbreg` command, however, estimates that parameter and provides unconditional estimates; see [R] **nbreg**.

5. `xtgee` with `corr(independent)` can be used to fit exponential regressions, but this requires specifying `scale(1)`. As with probit, the `xtgee`-reported standard errors will be only asymptotically equivalent to those produced by `streg, dist(exp) nohr` (see [ST] **streg**) because log is not the canonical link for the gamma family. `xtgee` cannot be used to fit exponential regressions on censored data.

   Using the `independent` correlation structure, the `xtgee` command will fit the same model as fitted with the `glm, irls` command if the family–link combination is the same.

6. If the `xtgee` command is equivalent to another command, then the use of `corr(independent)` and the `robust` option with `xtgee` corresponds to using both the `robust` option and the `cluster(`*varname*`)` option in the equivalent command, where *varname* corresponds to the i() group variable.

❑

`xtgee` is a direct generalization of the `glm, irls` command and will give the same output whenever the same family and link are specified together with an independent correlation structure. What makes `xtgee` useful is

1. the number of statistical models that it generalizes for use with panel data, many of which are not otherwise available in Stata;

2. the richer correlation structure `xtgee` allows even when models are available through other `xt` commands; and

3. the availability of robust standard errors (see [U] **23.14 Obtaining robust variance estimates**) even when the model and correlation structure are available through other `xt` commands.

In the following examples, we illustrate the relationships of `xtgee` with other Stata estimation commands. It is important to remember that although `xtgee` generalizes many other commands, the computational algorithm is different; therefore, the answers that you obtain will not be identical. The dataset we are using is a subset of the `nlswork` data (see [XT] **xt**); we are looking at observations prior to 1980.

▷ Example

We can use `xtgee` to perform ordinary least squares performed by `regress`:

```
. use http://www.stata-press.com/data/r8/nlswork2
(National Longitudinal Survey.  Young Women 14-26 years of age in 1968)

. generate age2 = age*age
(9 missing values generated)
```

```
. regress ln_w grade age age2
```

Source	SS	df	MS
Model	597.54468	3	199.18156
Residual	2265.74584	16081	.14089583
Total	2863.29052	16084	.178021047

```
Number of obs =   16085
F( 3, 16081) = 1413.68
Prob > F      =  0.0000
R-squared     =  0.2087
Adj R-squared =  0.2085
Root MSE      =  .37536
```

ln_wage	Coef.	Std. Err.	t	P>|t|	[95% Conf. Interval]	
grade	.0724483	.0014229	50.91	0.000	.0696592	.0752374
age	.1064874	.0083644	12.73	0.000	.0900922	.1228825
age2	-.0016931	.0001655	-10.23	0.000	-.0020174	-.0013688
_cons	-.8681487	.1024896	-8.47	0.000	-1.06904	-.6672577

```
. xtgee ln_w grade age age2, i(id) corr(indep) nmp

Iteration 1: tolerance = 1.310e-12

GEE population-averaged model
Group variable:                    idcode
Link:                            identity
Family:                          Gaussian
Correlation:                  independent

Scale parameter:                 .1408958

Pearson chi2(16081):              2265.75
Dispersion (Pearson):            .1408958
```

```
Number of obs      =     16085
Number of groups   =      3913
Obs per group: min =         1
               avg =       4.1
               max =         9
Wald chi2(3)       =   4241.04
Prob > chi2        =    0.0000

Deviance           =   2265.75
Dispersion         =  .1408958
```

ln_wage	Coef.	Std. Err.	z	P>|z|	[95% Conf. Interval]	
grade	.0724483	.0014229	50.91	0.000	.0696594	.0752372
age	.1064874	.0083644	12.73	0.000	.0900935	.1228812
age2	-.0016931	.0001655	-10.23	0.000	-.0020174	-.0013688
_cons	-.8681487	.1024896	-8.47	0.000	-1.069025	-.6672728

When the nmp option is specified, the coefficients and the standard errors produced by the estimators are exactly the same. Moreover, the scale parameter estimate from the xtgee command equals the MSE calculation from regress; both are estimates of the variance of the residuals.

◁

▷ Example

The identity link and Gaussian family produce regression-type models. With the independent correlation structure, we reproduce ordinary least squares. With the exchangeable correlation structure, we produce an equal-correlation linear regression estimator.

xtgee, fam(gauss) link(ident) corr(exch) is asymptotically equivalent to the weighted-GLS estimator provided by xtreg, re and to the full maximum-likelihood estimator provided by xtreg, mle. In balanced data, xtgee, fam(gauss) link(ident) corr(exch) and xtreg, mle produce exactly the same results. With unbalanced data, the results are close, but differ because the two estimators handle unbalanced data differently. For both balanced and unbalanced data, the results produced by xtgee, fam(gauss) link(ident) corr(exch) and xtreg, mle will differ from those produced by xtreg, re. Below, we demonstrate the use of the three estimators with unbalanced data. We begin with xtgee, then show the maximum likelihood estimator xtreg, mle, then show the GLS estimator xtreg, re, and finally show xtgee with the robust option.

```
. xtgee ln_w grade age age2, i(id) nolog
```

GEE population-averaged model   Number of obs   =   16085
Group variable: idcode Number of groups   =   3913
Link: identity Obs per group: min =   1
Family: Gaussian   avg =   4.1
Correlation: exchangeable   max =   9
                                       Wald chi2(3)   =   2918.26
Scale parameter: .1416586 Prob > chi2   =   0.0000

ln_wage	Coef.	Std. Err.	z	P>|z|	[95% Conf. Interval]	
grade	.0717731	.00211	34.02	0.000	.0676377	.0759086
age	.1077645	.006885	15.65	0.000	.0942701	.1212589
age2	-.0016381	.0001362	-12.03	0.000	-.001905	-.0013712
_cons	-.9480449	.0869277	-10.91	0.000	-1.11842	-.7776698

```
. xtreg ln_w grade age age2, i(id) mle
```

Fitting constant-only model:
Iteration 0:   log likelihood = -6035.2751
Iteration 1:   log likelihood = -5870.6718
Iteration 2:   log likelihood = -5858.9478
Iteration 3:   log likelihood = -5858.8244

Fitting full model:
Iteration 0:   log likelihood = -4591.9241
Iteration 1:   log likelihood = -4562.4406
Iteration 2:   log likelihood = -4562.3526

Random-effects ML regression   Number of obs   =   16085
Group variable (i): idcode   Number of groups   =   3913

Random effects u_i ~ Gaussian   Obs per group: min =   1
                                                     avg =   4.1
                                                     max =   9

                                       LR chi2(3)   =   2592.94
Log likelihood = -4562.3526   Prob > chi2   =   0.0000

ln_wage	Coef.	Std. Err.	z	P>|z|	[95% Conf. Interval]	
grade	.0717747	.0021419	33.51	0.000	.0675766	.0759728
age	.1077899	.0068265	15.79	0.000	.0944102	.1211696
age2	-.0016364	.000135	-12.12	0.000	-.0019011	-.0013718
_cons	-.9500833	.0863831	-11.00	0.000	-1.119391	-.7807755
/sigma_u	.2689639	.004085	65.84	0.000	.2609574	.2769704
/sigma_e	.2669944	.0017113	156.02	0.000	.2636404	.2703484
rho	.5036748	.0086443			.486734	.5206089

Likelihood-ratio test of sigma_u=0: chibar2(01)= 4996.22 Prob>=chibar2 = 0.000

(Continued on next page)

```
. xtreg ln_w grade age age2, i(id) re
```

| Random-effects GLS regression | Number of obs | = | 16085 |
| Group variable (i): idcode | Number of groups | = | 3913 |

R-sq:	within = 0.0983	Obs per group: min =	1
	between = 0.2946	avg =	4.1
	overall = 0.2076	max =	9

| Random effects u_i ~ Gaussian | Wald chi2(3) | = | 2875.09 |
| corr(u_i, X)   = 0 (assumed) | Prob > chi2 | = | 0.0000 |

ln_wage	Coef.	Std. Err.	z	P>|z|	[95% Conf. Interval]
grade	.0717757	.0021665	33.13	0.000	.0675295	.076022
age	.1078042	.0068126	15.82	0.000	.0944518	.1211566
age2	-.0016355	.0001347	-12.14	0.000	-.0018996	-.0013714
_cons	-.9512088	.0863141	-11.02	0.000	-1.120381	-.7820363

sigma_u	.27383336	
sigma_e	.2662536	
rho	.51403157	(fraction of variance due to u_i)

```
. xtgee ln_w grade age age2, i(id) nolog robust
```

GEE population-averaged model	Number of obs	=	16085
Group variable: idcode	Number of groups	=	3913
Link: identity	Obs per group: min =	1
Family: Gaussian	avg =	4.1
Correlation: exchangeable	max =	9
	Wald chi2(3)	=	2031.28
Scale parameter: .1416586	Prob > chi2	=	0.0000

(standard errors adjusted for clustering on idcode)

ln_wage	Coef.	Semi-robust Std. Err.	z	P>|z|	[95% Conf. Interval]
grade	.0717731	.0023341	30.75	0.000	.0671983	.0763479
age	.1077645	.0098097	10.99	0.000	.0885379	.1269911
age2	-.0016381	.0001964	-8.34	0.000	-.002023	-.0012532
_cons	-.9480449	.1195009	-7.93	0.000	-1.182262	-.7138274

In [R] **regress**, we noted the ability of regress, robust cluster() to produce inefficient coefficient estimates with valid standard errors for random-effects models. These standard errors are robust to model misspecification. The robust option of xtgee, on the other hand, requires that the model correctly specifies the mean.

◁

▷ Example

One of the features of xtgee is being able to estimate richer correlation structures. In the previous example, we fitted the model

```
. xtgee ln_w grade age age2, i(id)
```

After estimation, xtcorr will report the working correlation matrix **R**:

```
. xtcorr

Estimated within-idcode correlation matrix R:
         c1      c2      c3      c4      c5      c6      c7      c8      c9
r1   1.0000
r2   0.4851  1.0000
r3   0.4851  0.4851  1.0000
r4   0.4851  0.4851  0.4851  1.0000
r5   0.4851  0.4851  0.4851  0.4851  1.0000
r6   0.4851  0.4851  0.4851  0.4851  0.4851  1.0000
r7   0.4851  0.4851  0.4851  0.4851  0.4851  0.4851  1.0000
r8   0.4851  0.4851  0.4851  0.4851  0.4851  0.4851  0.4851  1.0000
r9   0.4851  0.4851  0.4851  0.4851  0.4851  0.4851  0.4851  0.4851  1.0000
```

The equal-correlation model corresponds to an exchangeable correlation structure, meaning that the correlation of observations within person is a constant. The working correlation estimated by xtgee is 0.4851. (xtreg, re, by comparison, reports .5140.) We constrained the model to have this simple correlation structure. What if we relaxed the constraint? To go to the other extreme, let's place no constraints on the matrix (other than it being symmetric). We do this by specifying correlation(unstructured), although we can abbreviate the option.

```
. xtgee ln_w grade age age2, i(id) t(year) corr(unstr) nolog

GEE population-averaged model                  Number of obs      =      16085
Group and time vars:          idcode year      Number of groups   =       3913
Link:                            identity       Obs per group: min =          1
Family:                          Gaussian                      avg =        4.1
Correlation:                 unstructured                      max =          9
                                                Wald chi2(3)       =    2405.20
Scale parameter:                  .1418513      Prob > chi2        =     0.0000
```

ln_wage	Coef.	Std. Err.	z	P>|z|	[95% Conf. Interval]	
grade	.0720684	.002151	33.50	0.000	.0678525	.0762843
age	.1008095	.0081471	12.37	0.000	.0848416	.1167775
age2	-.0015104	.0001617	-9.34	0.000	-.0018272	-.0011936
_cons	-.8645484	.1009488	-8.56	0.000	-1.062404	-.6666923

```
. xtcorr

Estimated within-idcode correlation matrix R:
         c1      c2      c3      c4      c5      c6      c7      c8      c9
r1   1.0000
r2   0.4355  1.0000
r3   0.4280  0.5597  1.0000
r4   0.3772  0.5012  0.5475  1.0000
r5   0.4031  0.5301  0.5027  0.6216  1.0000
r6   0.3664  0.4519  0.4783  0.5685  0.7306  1.0000
r7   0.2820  0.3606  0.3918  0.4012  0.4643  0.5022  1.0000
r8   0.3162  0.3446  0.4285  0.4389  0.4697  0.5223  0.6476  1.0000
r9   0.2149  0.3078  0.3337  0.3584  0.4866  0.4613  0.5791  0.7387  1.0000
```

This correlation matrix looks quite different from the previously constrained one, and shows, in particular, that the serial correlation of the residuals diminishes as the lag increases, although residuals separated by small lags are more correlated than, say, AR(1) would imply.

◁

▷ Example

In [XT] **xtprobit**, we showed a random-effects model of unionization using the union data described in [XT] **xt**. We performed the estimation using xtprobit, but said we could have used xtgee as well, and here we fit a population-averaged (equal-correlation) model for comparison:

```
. use http://www.stata-press.com/data/r8/union
(NLS Women 14-24 in 1968)

. xtgee union age grade not_smsa south southXt, i(id) fam(bin) link(probit)

Iteration 1: tolerance = .04796083
Iteration 2: tolerance = .00352657
Iteration 3: tolerance = .00017886
Iteration 4: tolerance = 8.654e-06
Iteration 5: tolerance = 4.150e-07
```

GEE population-averaged model				Number of obs	=	26200
Group variable:		idcode		Number of groups	=	4434
Link:		probit		Obs per group: min	=	1
Family:		binomial		avg	=	5.9
Correlation:		exchangeable		max	=	12
				Wald chi2(5)	=	241.66
Scale parameter:		1		Prob > chi2	=	0.0000

union	Coef.	Std. Err.	z	P>|z|	[95% Conf. Interval]	
age	.0031597	.0014678	2.15	0.031	.0002829	.0060366
grade	.0329992	.0062334	5.29	0.000	.020782	.0452163
not_smsa	-.0721799	.0275189	-2.62	0.009	-.1261159	-.0182439
south	-.409029	.0372213	-10.99	0.000	-.4819815	-.3360765
southXt	.0081828	.002545	3.22	0.001	.0031946	.0131709
_cons	-1.184799	.0890117	-13.31	0.000	-1.359259	-1.01034

Let us look at the correlation structure and then relax it:

```
. xtcorr
Estimated within-idcode correlation matrix R:

          c1      c2      c3      c4      c5      c6      c7      c8      c9
  r1  1.0000
  r2  0.4630  1.0000
  r3  0.4630  0.4630  1.0000
  r4  0.4630  0.4630  0.4630  1.0000
  r5  0.4630  0.4630  0.4630  0.4630  1.0000
  r6  0.4630  0.4630  0.4630  0.4630  0.4630  1.0000
  r7  0.4630  0.4630  0.4630  0.4630  0.4630  0.4630  1.0000
  r8  0.4630  0.4630  0.4630  0.4630  0.4630  0.4630  0.4630  1.0000
  r9  0.4630  0.4630  0.4630  0.4630  0.4630  0.4630  0.4630  0.4630  1.0000
 r10  0.4630  0.4630  0.4630  0.4630  0.4630  0.4630  0.4630  0.4630  0.4630
 r11  0.4630  0.4630  0.4630  0.4630  0.4630  0.4630  0.4630  0.4630  0.4630
 r12  0.4630  0.4630  0.4630  0.4630  0.4630  0.4630  0.4630  0.4630  0.4630
          c10     c11     c12
 r10  1.0000
 r11  0.4630  1.0000
 r12  0.4630  0.4630  1.0000
```

We estimate the fixed correlation between observations within person to be 0.4630. We have a lot of data (an average of 5.9 observations on 4,434 women), so estimating the full correlation matrix is feasible. Let's do that and then examine the results:

```
. xtgee union age grade not_smsa south southXt, i(id) t(t) fam(bin) link(probit)
> corr(unstr) nolog

GEE population-averaged model          Number of obs      =      26200
Group and time vars:          idcode t0  Number of groups    =       4434
Link:                           probit   Obs per group: min =          1
Family:                         binomial                 avg =        5.9
Correlation:                unstructured                 max =         12
                                         Wald chi2(5)       =     196.76
Scale parameter:                      1  Prob > chi2        =     0.0000
```

union	Coef.	Std. Err.	z	P>|z|	[95% Conf. Interval]	
age	.0020207	.0019768	1.02	0.307	-.0018539	.0058952
grade	.0349572	.0065627	5.33	0.000	.0220946	.0478198
not_smsa	-.0951058	.0291532	-3.26	0.001	-.152245	-.0379665
south	-.3891526	.0434868	-8.95	0.000	-.4743853	-.30392
southXt	.0078823	.0034032	2.32	0.021	.0012121	.0145524
_cons	-1.194276	.1000155	-11.94	0.000	-1.390303	-.9982495

```
. xtcorr

Estimated within-idcode correlation matrix R:

        c1      c2      c3      c4      c5      c6      c7      c8      c9
 r1  1.0000
 r2  0.6796  1.0000
 r3  0.6272  0.6628  1.0000
 r4  0.5365  0.5800  0.6170  1.0000
 r5  0.3377  0.3716  0.4037  0.4810  1.0000
 r6  0.3079  0.3771  0.4283  0.4591  0.6435  1.0000
 r7  0.3053  0.3630  0.3887  0.4299  0.4949  0.6407  1.0000
 r8  0.2807  0.3062  0.3251  0.3762  0.4691  0.5610  0.7000  1.0000
 r9  0.3045  0.3013  0.3042  0.3822  0.4620  0.5101  0.6093  0.6709  1.0000
r10  0.2324  0.2630  0.2779  0.3655  0.3987  0.4921  0.5878  0.5957  0.6308
r11  0.2369  0.2321  0.2716  0.3265  0.3555  0.4425  0.5094  0.5607  0.5740
r12  0.2400  0.2374  0.2561  0.3153  0.3478  0.3835  0.4782  0.4985  0.5404

        c10     c11     c12
r10  1.0000
r11  0.5706  1.0000
r12  0.5302  0.6406  1.0000
```

As before, we find that the correlation of residuals decreases as the lag increases, but more slowly than an AR(1) process, and then quickly decreases toward zero.

◁

▷ Example

In this example, we examine injury incidents among 20 airlines in each of 4 years. The data are fictional, and, as a matter of fact, are really from a random-effects model.

(Continued on next page)

```
. use http://www.stata-press.com/data/r8/airacc

. generate lnpm = ln(pmiles)

. xtgee i_cnt inprog, f(pois) i(airline) t(time) eform off(lnpm) nolog
```

GEE population-averaged model					Number of obs	=	80
Group variable:			airline		Number of groups	=	20
Link:			log		Obs per group: min	=	4
Family:			Poisson		avg	=	4.0
Correlation:			exchangeable		max	=	4
					Wald chi2(1)	=	5.27
Scale parameter:			1		Prob > chi2	=	0.0217

i_cnt	IRR	Std. Err.	z	P>|z|	[95% Conf. Interval]	
inprog	.9059936	.0389528	-2.30	0.022	.8327758	.9856487
lnpm	(offset)					

```
. xtcorr

Estimated within-airline correlation matrix R:

        c1      c2      c3      c4
r1  1.0000
r2  0.4606  1.0000
r3  0.4606  0.4606  1.0000
r4  0.4606  0.4606  0.4606  1.0000
```

Now, there are not really enough data here to reliably estimate the correlation without any constraints of structure, but here is what happens if we try:

```
. use http://www.stata-press.com/data/r8/airacc

. generate lnpm = ln(pmiles)

. xtgee i_cnt inprog, f(pois) i(airline) t(time) eform off(lnpm) corr(unstr) nolog
```

GEE population-averaged model					Number of obs	=	80
Group and time vars:			airline time		Number of groups	=	20
Link:			log		Obs per group: min	=	4
Family:			Poisson		avg	=	4.0
Correlation:			unstructured		max	=	4
					Wald chi2(1)	=	0.36
Scale parameter:			1		Prob > chi2	=	0.5496

i_cnt	IRR	Std. Err.	z	P>|z|	[95% Conf. Interval]	
inprog	.9791082	.0345486	-0.60	0.550	.9136826	1.049219
lnpm	(offset)					

```
. xtcorr

Estimated within-airline correlation matrix R:

        c1      c2      c3      c4
r1  1.0000
r2  0.5700  1.0000
r3  0.7164  0.4192  1.0000
r4  0.2383  0.3840  0.3521  1.0000
```

There is no sensible pattern to the correlations.

We admitted previously that we created this dataset from a random-effects Poisson model. We reran our data-creation program, and this time had it create 400 airlines rather than 20, still with 4 years of data each. Here is the equal-correlation model and estimated correlation structure,

```
. use http://www.stata-press.com/data/r8/airacc2

. xtgee i_cnt inprog, f(pois) i(airline) eform off(lnpm) nolog
GEE population-averaged model              Number of obs      =      1600
Group variable:                  airline   Number of groups   =       400
Link:                                log   Obs per group: min =         4
Family:                          Poisson                  avg =       4.0
Correlation:                 exchangeable                  max =         4
                                           Wald chi2(1)       =    111.80
Scale parameter:                       1   Prob > chi2        =    0.0000
```

i_cnt	IRR	Std. Err.	z	P>|z|	[95% Conf. Interval]	
inprog	.8915304	.0096807	-10.57	0.000	.8727571	.9107076
lnpm	(offset)					

```
. xtcorr

Estimated within-airline correlation matrix R:
         c1       c2       c3       c4
r1   1.0000
r2   0.5292   1.0000
r3   0.5292   0.5292   1.0000
r4   0.5292   0.5292   0.5292   1.0000
```

and here are the estimation results assuming unstructured correlation:

```
. use http://www.stata-press.com/data/r8/airacc2

. xtgee i_cnt inprog, f(pois) i(airline) corr(unstr) t(time) eform off(lnpm) nolog
GEE population-averaged model              Number of obs      =      1600
Group and time vars:        airline time   Number of groups   =       400
Link:                                log   Obs per group: min =         4
Family:                          Poisson                  avg =       4.0
Correlation:                 unstructured                  max =         4
                                           Wald chi2(1)       =    113.43
Scale parameter:                       1   Prob > chi2        =    0.0000
```

i_cnt	IRR	Std. Err.	z	P>|z|	[95% Conf. Interval]	
inprog	.8914155	.0096208	-10.65	0.000	.8727572	.9104728
lnpm	(offset)					

```
. xtcorr

Estimated within-airline correlation matrix R:
         c1       c2       c3       c4
r1   1.0000
r2   0.4733   1.0000
r3   0.5241   0.5749   1.0000
r4   0.5140   0.5049   0.5841   1.0000
```

The equal-correlation model estimated a fixed .5292, and above we have correlations ranging between .4733 and .5841 with little pattern in their structure.

◁

(Continued on next page)

Saved Results

xtgee saves in e():

Scalars

e(N)	number of observations	e(deviance)	deviance
e(N_g)	number of groups	e(chi2_dev)	χ^2 test of deviance
e(df_m)	model degrees of freedom	e(dispers)	deviance dispersion
e(g_max)	largest group size	e(chi2_dis)	χ^2 test of deviance dispersion
e(g_min)	smallest group size	e(tol)	target tolerance
e(g_avg)	average group size	e(dif)	achieved tolerance
e(chi2)	χ^2	e(phi)	scale parameter
e(df_pear)	degrees of freedom for Pearson χ^2		

Macros

e(cmd)	xtgee	e(tvar)	time variable
e(depvar)	name of dependent variable	e(vcetype)	covariance estimation method
e(family)	distribution family	e(chi2type)	Wald; type of model χ^2 test
e(link)	link function	e(offset)	offset
e(corr)	correlation structure	e(scorevars)	variable containing score
e(scale)	x2, dev, phi, or #; scale parameter	e(crittype)	optimization criterion
e(ivar)	variable denoting groups	e(predict)	program used to implement predict

Matrices

e(b)	coefficient vector	e(R)	estimated working correlation matrix
e(V)	variance–covariance matrix of the estimators		

Functions

e(sample)	marks estimation sample

Methods and Formulas

xtgee is implemented as an ado-file.

xtgee fits general linear models for panel data using the GEE approach described in Liang and Zeger (1986). Below we present the derivation of that estimator. A related method, referred to as GEE2, is described in Zhao and Prentice (1990) and Prentice and Zhao (1991). The GEE2 method attempts to gain efficiency in the estimation of $\boldsymbol{\beta}$ by specifying a parametric model for $\boldsymbol{\alpha}$, and then relies on assuming that the models for both the mean and dependency parameters are correct. Thus, there is a tradeoff in robustness for efficiency. The preliminary work of Liang, Zeger, and Qaqish (1987), however, indicates that there is little efficiency gained with this alternate approach.

In the GLM approach (see McCullagh and Nelder 1989), we assume that

$$h(\boldsymbol{\mu}_{i,j}) = x_{i,j}^{\mathrm{T}}\boldsymbol{\beta}$$
$$\mathrm{Var}(y_{i,j}) = g(\mu_{i,j})\phi$$
$$\boldsymbol{\mu}_i = E(\mathbf{y}_i) = \{h^{-1}(x_{i,1}^{\mathrm{T}}\boldsymbol{\beta}), \ldots, h^{-1}(x_{i,n_i}^{\mathrm{T}}\boldsymbol{\beta})\}^{\mathrm{T}}$$
$$\mathbf{A}_i = \mathrm{diag}\{g(\mu_{i,1}), \ldots, g(\mu_{i,n_i})\}$$
$$\mathrm{Cov}(\mathbf{y}_i) = \phi\mathbf{A}_i \quad \text{for independent observations.}$$

In the absence of a convenient likelihood function with which to work, one can rely on a multivariate analog of the quasi-score function introduced by Wedderburn (1974):

$$\mathbf{S}_{\boldsymbol{\beta}}(\boldsymbol{\beta}, \boldsymbol{\alpha}) = \sum_{i=1}^{m} \left(\frac{\partial \boldsymbol{\mu}_i}{\partial \boldsymbol{\beta}} \right)^{\mathrm{T}} \mathrm{Var}(\mathbf{y}_i)^{-1}(\mathbf{y}_i - \boldsymbol{\mu}_i) = 0$$

The correlation parameters $\boldsymbol{\alpha}$ can be solved for by simultaneously solving

$$\mathbf{S}_{\boldsymbol{\alpha}}(\boldsymbol{\beta}, \boldsymbol{\alpha}) = \sum_{i=1}^{m} \left(\frac{\partial \boldsymbol{\eta}_i}{\partial \boldsymbol{\alpha}} \right)^{\mathrm{T}} \mathbf{H}_i^{-1}(\mathbf{W}_i - \boldsymbol{\eta}_i) = 0$$

In the GEE approach to GLM, we let $\mathbf{R}_i(\boldsymbol{\alpha})$ be a "working" correlation matrix depending on the parameters in $\boldsymbol{\alpha}$ (see the *Correlation structures* section for the number of parameters), and we estimate $\boldsymbol{\beta}$ by solving the generalized estimating equation,

$$\mathbf{U}(\boldsymbol{\beta}) = \sum_{i=1}^{m} \frac{\partial \boldsymbol{\mu}_i}{\partial \boldsymbol{\beta}} \mathbf{V}_i^{-1}(\boldsymbol{\alpha})(\mathbf{y}_i - \boldsymbol{\mu}_i) = 0$$

$$\text{where} \quad \mathbf{V}_i(\boldsymbol{\alpha}) = \mathbf{A}_i^{1/2} \mathbf{R}_i(\boldsymbol{\alpha}) \mathbf{A}_i^{1/2}$$

To solve the above, we need only a crude approximation to the variance matrix. We can obtain one from a Taylor's series expansion, where

$$\mathrm{Cov}(\mathbf{y}_i) = \mathbf{L}_i \mathbf{Z}_i \mathbf{D}_i \mathbf{Z}_i^{\mathrm{T}} \mathbf{L}_i + \phi \mathbf{A}_i = \widetilde{\mathbf{V}}_i$$
$$\mathbf{L}_i = \mathrm{diag}\{\partial h^{-1}(u)/\partial u, u = x_{i,j}^{\mathrm{T}} \boldsymbol{\beta}, j = 1, \ldots, n_i\}$$

which allows that

$$\widehat{\mathbf{D}}_i \approx (\mathbf{Z}_i^{\mathrm{T}} \mathbf{Z}_i)^{-1} \mathbf{Z}_i \widehat{\mathbf{L}}_i^{-1} \left\{ (\mathbf{y}_i - \widehat{\boldsymbol{\mu}}_i)(\mathbf{y}_i - \widehat{\boldsymbol{\mu}}_i)^{\mathrm{T}} - \widehat{\phi} \widehat{\mathbf{A}}_i \right\} \widehat{\mathbf{L}}_i^{-1} \mathbf{Z}_i^{\mathrm{T}} (\mathbf{Z}_i' \mathbf{Z}_i)^{-1}$$
$$\widehat{\phi} = \sum_{i=1}^{m} \sum_{j=1}^{n_i} \frac{(y_{i,j} - \widehat{\mu}_{i,j})^2 - (\widehat{\mathbf{L}}_{i,j})^2 \mathbf{Z}_{i,j}^{\mathrm{T}} \widehat{\mathbf{D}}_i \mathbf{Z}_{i,j}}{g(\widehat{\mu}_{i,j})}$$

Calculation of GEE for GLM

Using the notation from Liang and Zeger (1986), let $\mathbf{y}_i = (y_{i,1}, \ldots, y_{i,n_i})^{\mathrm{T}}$ be the $n_i \times 1$ vector of outcome values, and let $\mathbf{X}_i = (x_{i,1}, \ldots, x_{i,n_i})^{\mathrm{T}}$ be the $n_i \times p$ matrix of covariate values for the ith subject $i = 1, \ldots, m$. We assume that the marginal density for $y_{i,j}$ may be written in exponential family notation as

$$f(y_{i,j}) = \exp\left[\{y_{i,j}\theta_{i,j} - a(\theta_{i,j}) + b(y_{i,j})\}\phi\right]$$

where $\theta_{i,j} = h(\eta_{i,j}), \eta_{i,j} = x_{i,j}\boldsymbol{\beta}$. Under this formulation, the first two moments are given by

$$E(y_{i,j}) = a'(\theta_{i,j}), \qquad \mathrm{Var}(y_{i,j}) = a''(\theta_{i,j})/\phi$$

We define the quantities (assuming that we have an $n \times n$ working correlation matrix $\mathbf{R}(\boldsymbol{\alpha})$),

$$\boldsymbol{\Delta}_i = \mathrm{diag}(d\theta_{i,j}/d\eta_{i,j}) \quad n \times n \text{ matrix}$$

$$\mathbf{A}_i = \mathrm{diag}\{a''(\theta_{i,j})\} \quad n \times n \text{ matrix}$$

$$\mathbf{S}_i = \mathbf{y}_i - a'(\boldsymbol{\theta}_i) \quad n \times 1 \text{ matrix}$$

$$\mathbf{D}_i = \mathbf{A}_i \boldsymbol{\Delta}_i \mathbf{X}_i \quad n \times p \text{ matrix}$$

$$\mathbf{V}_i = \mathbf{A}_i^{1/2} \mathbf{R}(\boldsymbol{\alpha}) \mathbf{A}_i^{1/2} \quad n \times n \text{ matrix}$$

such that the GEE becomes

$$\sum_{i=1}^{m} \mathbf{D}_i^{\mathrm{T}} \mathbf{V}_i^{-1} \mathbf{S}_i = 0$$

We then have that

$$\widehat{\boldsymbol{\beta}}_{j+1} = \widehat{\boldsymbol{\beta}}_j - \left\{ \sum_{i=1}^{m} \mathbf{D}_i^{\mathrm{T}}(\widehat{\boldsymbol{\beta}}_j) \widetilde{\mathbf{V}}_i^{-1}(\widehat{\boldsymbol{\beta}}_j) \mathbf{D}_i(\widehat{\boldsymbol{\beta}}_j) \right\}^{-1} \left\{ \sum_{i=1}^{m} \mathbf{D}_i^{\mathrm{T}}(\widehat{\boldsymbol{\beta}}_j) \widetilde{\mathbf{V}}_i^{-1}(\widehat{\boldsymbol{\beta}}_j) \mathbf{S}_i(\widehat{\boldsymbol{\beta}}_j) \right\}$$

where the term

$$\left\{ \sum_{i=1}^{m} \mathbf{D}_i^{\mathrm{T}}(\widehat{\boldsymbol{\beta}}_j) \widetilde{\mathbf{V}}_i^{-1}(\widehat{\boldsymbol{\beta}}_j) \mathbf{D}_i(\widehat{\boldsymbol{\beta}}_j) \right\}^{-1}$$

is what we call the IRLS variance estimate (iteratively reweighted least squares). It is used to calculate the standard errors if the `robust` option is not specified. See Liang and Zeger (1986) for the calculation of the robust variance estimator.

Define

$$\mathbf{D} = (\mathbf{D}_1^{\mathrm{T}}, \ldots, \mathbf{D}_m^{\mathrm{T}})$$

$$\mathbf{S} = (\mathbf{S}_1^{\mathrm{T}}, \ldots, \mathbf{S}_m^{\mathrm{T}})^{\mathrm{T}}$$

$$\widetilde{\mathbf{V}} = nm \times nm \text{ block diagonal matrix with } \widetilde{\mathbf{V}}_i$$

$$\mathbf{Z} = \mathbf{D}\boldsymbol{\beta} - \mathbf{S}$$

At a given iteration, the correlation parameters $\boldsymbol{\alpha}$ and scale parameter ϕ can be estimated from the current Pearson residuals, defined by

$$\widehat{r}_{i,j} = \{y_{i,j} - a'(\widehat{\theta}_{i,j})\}/\{a''(\widehat{\theta}_{i,j})\}^{1/2}$$

where $\widehat{\theta}_{i,j}$ depends on the current value for $\widehat{\boldsymbol{\beta}}$. We can then estimate ϕ by

$$\widehat{\phi}^{-1} = \sum_{i=1}^{m} \sum_{j=1}^{n_i} \widehat{r}_{i,j}^2 / (N - p)$$

As the above general derivation is complicated, let's follow the derivation of the Gaussian family with the identity link (regression) to illustrate the generalization. After making appropriate substitutions, we will see a familiar updating equation. First, we rewrite the updating equation for $\boldsymbol{\beta}$ as

$$\widehat{\boldsymbol{\beta}}_{j+1} = \widehat{\boldsymbol{\beta}}_j - \mathbf{Z}_1^{-1} \mathbf{Z}_2$$

and then derive for $\mathbf{Z}_1$ and $\mathbf{Z}_2$.

$$\mathbf{Z}_1 = \sum_{i=1}^{m} \mathbf{D}_i^{\mathrm{T}}(\widehat{\boldsymbol{\beta}}_j) \widetilde{\mathbf{V}}_i^{-1}(\widehat{\boldsymbol{\beta}}_j) \mathbf{D}_i(\widehat{\boldsymbol{\beta}}_j) = \sum_{i=1}^{m} \mathbf{X}_i^{\mathrm{T}} \boldsymbol{\Delta}_i^{\mathrm{T}} \mathbf{A}_i^{\mathrm{T}} \{ \mathbf{A}_i^{1/2} \mathbf{R}(\boldsymbol{\alpha}) \mathbf{A}_i^{1/2} \}^{-1} \mathbf{A}_i \boldsymbol{\Delta}_i \mathbf{X}_i$$

$$= \sum_{i=1}^{m} \mathbf{X}_i^{\mathrm{T}} \operatorname{diag} \left\{ \frac{\partial \theta_{i,j}}{\partial (\mathbf{X}\boldsymbol{\beta})} \right\} \operatorname{diag} \{ a''(\theta_{i,j}) \} \left[\operatorname{diag} \{ a''(\theta_{i,j}) \}^{1/2} \mathbf{R}(\boldsymbol{\alpha}) \operatorname{diag} \{ a''(\theta_{i,j}) \}^{1/2} \right]^{-1}$$

$$\operatorname{diag} \{ a''(\theta_{i,j}) \} \operatorname{diag} \left\{ \frac{\partial \theta_{i,j}}{\partial (\mathbf{X}\boldsymbol{\beta})} \right\} \mathbf{X}_i$$

$$= \sum_{i=1}^{m} \mathbf{X}_i^{\mathrm{T}} \mathbf{II} (\mathbf{III})^{-1} \mathbf{II} \mathbf{X}_i = \sum_{i=1}^{m} \mathbf{X}_i^{\mathrm{T}} \mathbf{X}_i = \mathbf{X}^{\mathrm{T}} \mathbf{X}$$

$$\mathbf{Z}_2 = \sum_{i=1}^{m} \mathbf{D}_i^{\mathrm{T}}(\widehat{\boldsymbol{\beta}}_j) \widetilde{\mathbf{V}}_i^{-1}(\widehat{\boldsymbol{\beta}}_j) \mathbf{S}_i(\widehat{\boldsymbol{\beta}}_j) = \sum_{i=1}^{m} \mathbf{X}_i^{\mathrm{T}} \boldsymbol{\Delta}_i^{\mathrm{T}} \mathbf{A}_i^{\mathrm{T}} \{ \mathbf{A}_i^{1/2} \mathbf{R}(\boldsymbol{\alpha}) \mathbf{A}_i^{1/2} \}^{-1} \left(\mathbf{y}_i - \mathbf{X}_i \widehat{\boldsymbol{\beta}}_j \right)$$

$$= \sum_{i=1}^{m} \mathbf{X}_i^{\mathrm{T}} \operatorname{diag} \left\{ \frac{\partial \theta_{i,j}}{\partial (\mathbf{X}\boldsymbol{\beta})} \right\} \operatorname{diag} \{ a''(\theta_{i,j}) \} \left[\operatorname{diag} \{ a''(\theta_{i,j}) \}^{1/2} \mathbf{R}(\boldsymbol{\alpha}) \operatorname{diag} \{ a''(\theta_{i,j}) \}^{1/2} \right]^{-1}$$

$$\left(\mathbf{y}_i - \mathbf{X}_i \widehat{\boldsymbol{\beta}}_j \right)$$

$$= \sum_{i=1}^{m} \mathbf{X}_i \mathbf{II} (\mathbf{III})^{-1} (\mathbf{y}_i - \mathbf{X}_i \widehat{\boldsymbol{\beta}}_j) = \sum_{i=1}^{m} \mathbf{X}_i^{\mathrm{T}} (\mathbf{y}_i - \mathbf{X}_i \widehat{\boldsymbol{\beta}}_j) = \mathbf{X}^{\mathrm{T}} \widehat{s}_j$$

So, that means that we may write the update formula as

$$\widehat{\boldsymbol{\beta}}_{j+1} = \widehat{\boldsymbol{\beta}}_j - (\mathbf{X}^{\mathrm{T}} \mathbf{X})^{-1} \mathbf{X}^{\mathrm{T}} \widehat{s}_j$$

which is the same formula for IRLS in regression.

Correlation structures

The working correlation matrix $\mathbf{R}$ is a function of $\boldsymbol{\alpha}$, and is more accurately written as $\mathbf{R}(\boldsymbol{\alpha})$. Depending on the assumed correlation structure, $\boldsymbol{\alpha}$ might be

Independent	No parameters to estimate
Exchangeable	$\boldsymbol{\alpha}$ is a scalar
Autoregressive	$\boldsymbol{\alpha}$ is a vector
Stationary	$\boldsymbol{\alpha}$ is a vector
Nonstationary	$\boldsymbol{\alpha}$ is a matrix
Unstructured	$\boldsymbol{\alpha}$ is a matrix

Also note that throughout the estimation of a general unbalanced panel, it is more proper to discuss $\mathbf{R}_i$, which is the upper left $n_i \times n_i$ submatrix of the ultimately saved matrix in e(R), which is $\max\{n_i\} \times \max\{n_i\}$.

The only panels that enter into the estimation for a lag-dependent correlation structure are those with $n_i > g$ (assuming a lag of g). xtgee drops panels with too few observations (and mentions when it does so).

Independent

The working correlation matrix $\mathbf{R}$ is an identity matrix.

Exchangeable

$$\alpha = \sum_{i=1}^{m} \left\{ \frac{\sum_{j=1}^{n_i} \sum_{k=1}^{n_i} \widehat{r}_{i,j} \widehat{r}_{i,k} - \sum_{j=1}^{n_i} \widehat{r}_{i,j}^2}{n_i(n_i - 1)} \right\} \Bigg/ \left(\sum_{i=1}^{m} \frac{\sum_{j=1}^{n_i} \widehat{r}_{i,j}^2}{n_i} \right)$$

and the working correlation matrix is given by

$$\mathbf{R}_{s,t} = \begin{cases} 1 & s = t \\ \alpha & \text{otherwise} \end{cases}$$

Autoregressive and stationary

These two structures require g parameters to be estimated so that α is a vector of length $g + 1$ (the first element of α is 1).

$$\alpha = \sum_{i=1}^{m} \left(\frac{\sum_{j=1}^{n_i} \widehat{r}_{i,j}^2}{n_i} , \frac{\sum_{j=1}^{n_i-1} \widehat{r}_{i,j} \widehat{r}_{i,j+1}}{n_i} , \cdots , \frac{\sum_{j=1}^{n_i-g} \widehat{r}_{i,j} \widehat{r}_{i,j+g}}{n_i} \right) \Bigg/ \left(\sum_{i=1}^{m} \frac{\sum_{j=1}^{n_i} \widehat{r}_{i,j}^2}{n_i} \right)$$

The working correlation matrix for the AR model is calculated as a function of Toeplitz matrices formed from the α vector. See Newton (1988) for a discussion of the algorithm. The working correlation matrix for the stationary model is given by

$$\mathbf{R}_{s,t} = \begin{cases} \alpha_{1,|s-t|} & \text{if } |s - t| \leq g \\ 0 & \text{otherwise} \end{cases}$$

Nonstationary and Unstructured

These two correlation structures require a matrix of parameters. α is estimated (where we replace $\widehat{r}_{i,j} = 0$ whenever $i > n_i$ or $j > n_i$) as

$$\alpha = \sum_{i=1}^{m} m \begin{pmatrix} N_{1,1}^{-1} \widehat{r}_{i,1}^2 & N_{1,2}^{-1} \widehat{r}_{i,1} \widehat{r}_{i,2} & \cdots & N_{1,n}^{-1} \widehat{r}_{i,1} \widehat{r}_{i,n} \\ N_{2,1}^{-1} \widehat{r}_{i,2} \widehat{r}_{i,1} & N_{2,2}^{-1} \widehat{r}_{i,2}^2 & \cdots & N_{2,n}^{-1} \widehat{r}_{i,2} \widehat{r}_{i,n} \\ \vdots & \vdots & \ddots & \vdots \\ N_{n,1}^{-1} \widehat{r}_{i,n_i} \widehat{r}_{i,1} & N_{n,2}^{-1} \widehat{r}_{i,n_i} \widehat{r}_{i,2} & \cdots & N_{n,n}^{-1} \widehat{r}_{i,n}^2 \end{pmatrix} \Bigg/ \left(\sum_{i=1}^{m} \frac{\sum_{j=1}^{n_i} \widehat{r}_{i,j}^2}{n_i} \right)$$

where $N_{p,q} = \sum_{i=1}^{m} I(i,p,q)$ and

$$I(i,p,q) = \begin{cases} 1 & \text{if panel } i \text{ has valid observations at times p and q} \\ 0 & \text{otherwise} \end{cases}$$

where $N_{i,j} = \min(N_i, N_j)$, N_i = number of panels observed at time i, and $n = \max(n_1, n_2, \ldots, n_m)$.

The working correlation matrix for the nonstationary model is given by

$$\mathbf{R}_{s,t} = \begin{cases} 1 & \text{if } s = t \\ \alpha_{s,t} & \text{if } 0 < |s - t| \leq g \\ 0 & \text{otherwise} \end{cases}$$

The working correlation matrix for the unstructured model is given by

$$\mathbf{R}_{s,t} = \begin{cases} 1 & \text{if } s = t \\ \boldsymbol{\alpha}_{s,t} & \text{otherwise} \end{cases}$$

such that the unstructured model is equal to the nonstationary model at lag $g = n - 1$, where the panels are balanced with $n_i = n$ for all i.

References

Hardin, J. W. 2002. The robust variance estimator for two-stage models. *The Stata Journal* 2: 253–266.

Hardin, J. W. and J. M. Hilbe. 2003. *Generalized Estimating Equations.* Boca Raton, FL: Chapman & Hall/CRC.

Hosmer, D. W., Jr., and S. Lemeshow. 2002. *Applied Logistic Regression.* 2d ed. New York: John Wiley & Sons.

Kleinbaum, D. G. and M. Klein. 2002. *Logistic Regression: A Self-Learning Text.* 2d ed. New York: Springer.

Liang, K.-Y. 1987. Estimating functions and approximate conditional likelihood. *Biometrika* 4: 695–702.

Liang, K.-Y. and S. L. Zeger. 1986. Longitudinal data analysis using generalized linear models. *Biometrika* 73: 13–22.

Liang, K.-Y., S. L. Zeger, and B. Qaqish. 1987. Multivariate regression analyses for categorical data. *Journal of the Royal Statistical Society*, Series B 54: 3–40.

McCullagh, P. and J. A. Nelder. 1989. *Generalized Linear Models.* 2d ed. London: Chapman & Hall.

Nelder, J. A. and R. W. M. Wedderburn. 1972. Generalized linear models. *Journal of the Royal Statistical Society*, Series A 135: 370–384.

Newton, H. J. 1988. *TIMESLAB: A Time Series Analysis Laboratory*, Belmont, CA: Wadsworth & Brooks/Cole.

Pendergast, J. F., S. J. Gange, M. A. Newton, M. J. Lindstrom, M. Palta, and M. R. Fisher. 1996. A survey of methods for analyzing clustered binary response data. *International Statistical Review* 64: 89–118.

Prentice, R. L. and L. P. Zhao. 1991. Estimating equations for parameters in means and covariances of multivariate discrete and continuous responses. *Biometrics* 47: 825–839.

Rabe-Hesketh, S., A. Pickles, and C. Taylor. 2000. sg129: Generalized linear latent and mixed models. *Stata Technical Bulletin* 53: 47–57. Reprinted in *Stata Technical Bulletin Reprints*, vol. 9, pp. 293–307.

Rabe-Hesketh, S., A. Skrondal, and A. Pickles. 2002 Reliable estimation of generalized linear mixed models using adaptive quadrature. *The Stata Journal* 2: 1–21.

Wedderburn, R. W. M. 1974. Quasi-likelihood functions, generalized linear models, and the Gauss–Newton method. *Biometrika* 61: 439–447.

Zeger, S. L. and K.-Y. Liang. 1986. Longitudinal data analysis for discrete and continuous outcomes. *Biometrics* 42: 121–130.

Zeger, S. L., K.-Y. Liang, and P. S. Albert. 1988. Models for longitudinal data: a generalized estimating equation approach *Biometrics* 44: 1049–1060.

Zhao, L. P. and R. L. Prentice. 1990. Correlated binary regression using a quadratic exponential model. *Biometrika* 77: 642–648.

(Continued on next page)

Also See

Complementary:	[XT] **xtdata**, [XT] **xtdes**, [XT] **xtsum**, [XT] **xttab**,
	[R] **adjust**, [R] **lincom**, [R] **mfx**, [R] **nlcom**, [R] **predict**, [R] **predictnl**,
	[R] **suest**, [R] **test**, [R] **testnl**, [R] **vce**
Related:	[XT] **xtcloglog**, [XT] **xtgls**, [XT] **xtintreg**, [XT] **xtlogit**, [XT] **xtnbreg**,
	[XT] **xtpcse**, [XT] **xtpoisson**, [XT] **xtprobit**, [XT] **xtreg**, [XT] **xtregar**,
	[XT] **xttobit**,
	[R] **logistic**, [R] **regress**,
	[SVY] **svy estimators**,
	[TS] **prais**
Background:	[U] **16.5 Accessing coefficients and standard errors**,
	[U] **23 Estimation and post-estimation commands**,
	[U] **23.14 Obtaining robust variance estimates**,
	[U] **23.15 Obtaining scores**,
	[XT] **xt**

Title

> **xtgls** — Fit panel-data models using GLS

Syntax

> xtgls *depvar* [*varlist*] [*weight*] [if *exp*] [in *range*] [, i(*varname*) t(*varname*)
>
> force igls nmk panels(iid|heteroskedastic|correlated)
>
> corr(independent|ar1|psar1) noconstant level(#) tolerance(#)
>
> iterate(#) nolog rhotype(regress|dw|freg|nagar|theil|tscorr)]

by ... : may be used with xtgls; see [R] **by**.
aweights are allowed; see [U] **14.1.6 weight**.
xtgls shares the features of all estimation commands; see [U] **23 Estimation and post-estimation commands**.

Syntax for predict

> predict [*type*] *newvarname* [if *exp*] [in *range*] [, [xb | stdp]]

These statistics are available both in and out of sample; type predict ... if e(sample) ... if wanted only for
the estimation sample.

Description

xtgls fits cross-sectional time-series linear models using feasible generalized least squares. This
command allows estimation in the presence of AR(1) autocorrelation within panels and cross-sectional
correlation and/or heteroskedasticity across panels.

Options

i(*varname*) specifies the variable that identifies the panel to which the observation belongs. You can
specify the i() option the first time you estimate, or you can use the iis command to set i()
beforehand. Note that it is not necessary to specify i() if the data have been previously tsset, or
if iis has been previously specified—in these cases, the group variable is taken from the previous
setting. See [XT] **xt**.

t(*varname*) specifies the variable that contains the time at which the observation was made. You
can specify the t() option the first time you estimate, or you can use the tis command to set
t() beforehand. Note that it is not necessary to specify t() if the data have been previously
tsset, or if tis has been previously specified—in these cases, the group variable is taken from
the previous setting. See [XT] **xt**.

xtgls does not need to know t() in all cases, and, in those cases, whether you specify t()
makes no difference. We note in the descriptions of the panels() and corr() options when t()
is required. When t() is required, it is also required that the observations be spaced equally over
time; however, see option force below.

85

force specifies that estimation is to be forced even though t() is not equally spaced. This is relevant only for correlation structures that require knowledge of t(). These correlation structures require that observations be equally spaced so that calculations based on lags correspond to a constant time change. If you specify a t() variable that indicates observations are not equally spaced, xtgls will refuse to fit the (time-dependent) model. If you also specify force, xtgls will fit the model and assume that the lags based on the data ordered by t() are appropriate.

igls requests an iterated GLS estimator instead of the two-step GLS estimator in the case of a nonautocorrelated model, or instead of the three-step GLS estimator in the case of an autocorrelated model. The iterated GLS estimator converges to the MLE for the corr(independent) models, but does not for the other corr() models.

nmk specifies standard errors are to be normalized by $n - k$, where k is the number of parameters estimated, rather than n, the number of observations. The one used varies among authors. Greene (2003, 322) recommends n, and remarks that whether you use n or $n - k$ does not make the variance calculation unbiased in these models.

panels(*pdist*) specifies the error structure across panels.

panels(iid) specifies a homoskedastic error structure with no cross-sectional correlation. This is the default.

panels(heteroskedastic) (typically abbreviated p(h)) specifies a heteroskedastic error structure with no cross-sectional correlation.

panels(correlated) (abbreviation p(c)) specifies heteroskedastic error structure with cross-sectional correlation. If p(c) is specified, you must also specify t(). Note that the results will be based on a generalized inverse of a singular matrix unless $T \geq m$ (the number of time periods is greater than or equal to the number of panels).

corr(*corr*) specifies the assumed autocorrelation within panels.

corr(independent) (abbreviation c(i)) specifies that there is no autocorrelation. This is the default.

corr(ar1) (abbreviation c(a)) specifies that, within panels, there is AR(1) autocorrelation, and that the coefficient of the AR(1) process is common to all the panels. If c(ar1) is specified, you must also specify t().

corr(psar1) (abbreviation c(p)) specifies that, within panels, there is AR(1) autocorrelation, and that the coefficient of the AR(1) process is specific to each panel. psar1 stands for panel-specific AR(1). If c(psar1) is specified, t() must also be specified.

noconstant suppresses the estimation of the constant term (intercept).

level(*#*) specifies the confidence level, in percent, for confidence intervals. The default is level(95) or as set by set level; see [U] **23.6 Specifying the width of confidence intervals**.

tolerance(*#*) specifies the convergence criterion for the maximum change in the estimated coefficient vector between iterations; tol(1e-6) is the default.

iterate(*#*) specifies the maximum number of iterations allowed in fitting the model; iterate(100) is the default. You should never have to specify this option.

nolog suppresses the iteration log.

rhotype(*calc*) specifies the method to be used to calculate the autocorrelation parameter. Allowed strings for *calc* are

regress	regression using lags (the default)
dw	Durbin–Watson calculation
freg	regression using leads
nagar	Nagar calculation
theil	Theil calculation
tscorr	time-series autocorrelation calculation

All the calculations are asymptotically equivalent and all are consistent; this is a rarely used option.

Options for predict

xb, the default, calculates the linear prediction.

stdp calculates the standard error of the linear prediction.

Remarks

Remarks are presented under the headings

Introduction
Heteroskedasticity across panels
Correlation across panels (cross-sectional correlation)
Autocorrelation within panels

Introduction

Information on GLS can be found in Greene (2003), Maddala (1992), Davidson and MacKinnon (1993), and Judge et al. (1985).

If you have a large number of panels relative to time periods, see [XT] **xtreg** and [XT] **xtgee**. xtgee, in particular, provides similar capabilities as xtgls, but does not allow cross-sectional correlation. On the other hand, xtgee will allow a richer description of the correlation within panels subject to the constraint that the same correlations apply to all panels. That is, xtgls provides two unique features:

1. Cross-sectional correlation may be modeled (panels(correlated)).

2. Within panels, the AR(1) correlation coefficient may be unique (corr(psar1)).

It is also true that xtgls allows models with heteroskedasticity and no cross-sectional correlation, whereas, strictly speaking, xtgee does not, but xtgee with the robust option relaxes the assumption of equal variances, at least as far as the standard error calculation is concerned.

In addition, xtgls, panels(iid) corr(independent) nmk is equivalent to regress.

The nmk option uses $n - k$ rather than n to normalize the variance calculation.

To fit a model with autocorrelated errors (corr(ar1) or corr(psar1)), the data must be equally spaced in time. To fit a model with cross-sectional correlation (panels(correlated)), panels must have the same number of observations (be balanced).

The equation from which the models are developed is given by

$$y_{it} = \mathbf{x}_{it}\boldsymbol{\beta} + \epsilon_{it}$$

where $i = 1, \ldots, m$ is the number of units (or panels) and $t = 1, \ldots, T_i$ is the number of observations for panel i. This model can equally be written as

$$
\begin{bmatrix} \mathbf{y}_1 \\ \mathbf{y}_2 \\ \vdots \\ \mathbf{y}_m \end{bmatrix} = \begin{bmatrix} \mathbf{X}_1 \\ \mathbf{X}_2 \\ \vdots \\ \mathbf{X}_m \end{bmatrix} \boldsymbol{\beta} + \begin{bmatrix} \boldsymbol{\epsilon}_1 \\ \boldsymbol{\epsilon}_2 \\ \vdots \\ \boldsymbol{\epsilon}_m \end{bmatrix}
$$

The variance matrix of the disturbance terms can be written as

$$
E[\boldsymbol{\epsilon}\boldsymbol{\epsilon}'] = \boldsymbol{\Omega} = \begin{bmatrix} \sigma_{1,1}\boldsymbol{\Omega}_{1,1} & \sigma_{1,2}\boldsymbol{\Omega}_{1,2} & \cdots & \sigma_{1,m}\boldsymbol{\Omega}_{1,m} \\ \sigma_{2,1}\boldsymbol{\Omega}_{2,1} & \sigma_{2,2}\boldsymbol{\Omega}_{2,2} & \cdots & \sigma_{2,m}\boldsymbol{\Omega}_{2,m} \\ \vdots & \vdots & \ddots & \vdots \\ \sigma_{m,1}\boldsymbol{\Omega}_{m,1} & \sigma_{m,2}\boldsymbol{\Omega}_{m,2} & \cdots & \sigma_{m,m}\boldsymbol{\Omega}_{m,m} \end{bmatrix}
$$

In order for the $\boldsymbol{\Omega}_{i,j}$ matrices to be parameterized to model cross-sectional correlation, they must be square (balanced panels).

In these models, we assume that the coefficient vector $\boldsymbol{\beta}$ is the same for all panels, and consider a variety of models by changing the assumptions on the structure of $\boldsymbol{\Omega}$.

For the classic OLS regression model, we have

$$
E[\epsilon_{i,t}] = 0
$$
$$
\text{Var}[\epsilon_{i,t}] = \sigma^2
$$
$$
\text{Cov}[\epsilon_{i,t}, \epsilon_{j,s}] = 0 \qquad \text{if } t \neq s \text{ or } i \neq j
$$

This amounts to assuming that $\boldsymbol{\Omega}$ has the structure given by

$$
\boldsymbol{\Omega} = \begin{bmatrix} \sigma^2\mathbf{I} & \mathbf{0} & \cdots & \mathbf{0} \\ \mathbf{0} & \sigma^2\mathbf{I} & \cdots & \mathbf{0} \\ \vdots & \vdots & \ddots & \vdots \\ \mathbf{0} & \mathbf{0} & \cdots & \sigma^2\mathbf{I} \end{bmatrix}
$$

whether or not the panels are balanced (the $\mathbf{0}$ matrices may be rectangular). The classic OLS assumptions are the default panels(uncorrelated) and corr(independent) options for this command.

Heteroskedasticity across panels

In many cross-sectional datasets, the variance for each of the panels will differ. It is common to have data on countries, states, or other units that have variation of scale. The heteroskedastic model is specified by including the panels(heteroskedastic) option, which assumes that

$$\Omega = \begin{bmatrix} \sigma_1^2 \mathbf{I} & 0 & \cdots & 0 \\ 0 & \sigma_2^2 \mathbf{I} & \cdots & 0 \\ \vdots & \vdots & \ddots & \vdots \\ 0 & 0 & \cdots & \sigma_m^2 \mathbf{I} \end{bmatrix}$$

▷ Example

Greene (2003, 329 and *http://www.prenhall.com/greene*) reprints data in a classic study of investment demand by Grunfeld and Griliches (1960). Below, we allow the variances to differ for each of the five companies.

```
. use http://www.stata-press.com/data/r8/invest2

. xtgls invest market stock, i(company) panels(hetero)

Cross-sectional time-series FGLS regression

Coefficients:  generalized least squares
Panels:        heteroskedastic
Correlation:   no autocorrelation

Estimated covariances      =          5      Number of obs      =         100
Estimated autocorrelations =          0      Number of groups   =           5
Estimated coefficients     =          3      Time periods       =          20
                                             Wald chi2(2)       =      865.38
Log likelihood             = -570.1305       Prob > chi2        =      0.0000
```

invest	Coef.	Std. Err.	z	P>|z|	[95% Conf. Interval]	
market	.0949905	.007409	12.82	0.000	.0804692	.1095118
stock	.3378129	.0302254	11.18	0.000	.2785722	.3970535
_cons	-36.2537	6.124363	-5.92	0.000	-48.25723	-24.25017

◁

Correlation across panels (cross-sectional correlation)

We may wish to assume that the error terms of panels are correlated, in addition to having different scale variances. This variance structure is specified by including the panels(correlated) option and is given by

$$\Omega = \begin{bmatrix} \sigma_1^2 \mathbf{I} & \sigma_{1,2} \mathbf{I} & \cdots & \sigma_{1,m} \mathbf{I} \\ \sigma_{2,1} \mathbf{I} & \sigma_2^2 \mathbf{I} & \cdots & \sigma_{2,m} \mathbf{I} \\ \vdots & \vdots & \ddots & \vdots \\ \sigma_{m,1} \mathbf{I} & \sigma_{m,2} \mathbf{I} & \cdots & \sigma_m^2 \mathbf{I} \end{bmatrix}$$

Note that since we must estimate cross-sectional correlation in this model, the panels must be balanced (and $T \geq m$ for valid results). In addition, we must now specify the t() option so that xtgls knows how the observations within panels are ordered.

▷ Example

```
. xtgls invest market stock, i(company) t(time) panels(correlated)
Cross-sectional time-series FGLS regression
Coefficients:  generalized least squares
Panels:        heteroskedastic with cross-sectional correlation
Correlation:   no autocorrelation
Estimated covariances      =          15          Number of obs     =         100
Estimated autocorrelations =           0          Number of groups  =           5
Estimated coefficients     =           3          Time periods      =          20
                                                  Wald chi2(2)      =     1285.19
Log likelihood               = -537.8045          Prob > chi2       =      0.0000
```

invest	Coef.	Std. Err.	z	P>|z|	[95% Conf. Interval]	
market	.0961894	.0054752	17.57	0.000	.0854583	.1069206
stock	.3095321	.0179851	17.21	0.000	.2742819	.3447822
_cons	-38.36128	5.344871	-7.18	0.000	-48.83703	-27.88552

The estimated cross-sectional covariances are stored in e(Sigma).

```
. matrix list e(Sigma)
symmetric e(Sigma)[5,5]
             _ee        _ee2        _ee3        _ee4        _ee5
  _ee   9410.9061
 _ee2  -168.04631    755.85077
 _ee3  -1915.9538   -4163.3434    34288.49
 _ee4  -1129.2896   -80.381742   2259.3242   633.42367
 _ee5   258.50132    4035.872   -27898.235  -1170.6801   33455.511
```

◁

▷ Example

We can obtain the MLE results by specifying the igls option, which iterates the GLS estimation technique to convergence:

```
. xtgls invest market stock, i(company) t(time) panels(correlated) igls
Iteration 1:  tolerance = .2127384
Iteration 2:  tolerance = .22817
  (output omitted )
Iteration 1046:  tolerance = 1.000e-07

Cross-sectional time-series FGLS regression
Coefficients:  generalized least squares
Panels:        heteroskedastic with cross-sectional correlation
Correlation:   no autocorrelation
Estimated covariances      =          15          Number of obs     =         100
Estimated autocorrelations =           0          Number of groups  =           5
Estimated coefficients     =           3          Time periods      =          20
                                                  Wald chi2(2)      =      558.51
Log likelihood               = -515.4222          Prob > chi2       =      0.0000
```

invest	Coef.	Std. Err.	z	P>|z|	[95% Conf. Interval]	
market	.023631	.004291	5.51	0.000	.0152207	.0320413
stock	.1709472	.0152526	11.21	0.000	.1410526	.2008417
_cons	-2.216508	1.958845	-1.13	0.258	-6.055774	1.622759

◁

Autocorrelation within panels

The individual identity matrices along the diagonal of Ω may be replaced with more general structures in order to allow for serial correlation. xtgls allows three options so that you may assume a structure with corr(independent) (no autocorrelation); corr(ar1) (serial correlation where the correlation parameter is common for all panels); or corr(psar1) (serial correlation where the correlation parameter is unique for each panel).

The restriction of a common autocorrelation parameter is reasonable when the individual correlations are nearly equal and the time series are short.

If the restriction of a common autocorrelation parameter is reasonable, this allows us to use more information in estimating the autocorrelation parameter, and thus to produce a more reasonable estimate of the regression coefficients.

▷ **Example**

If corr(ar1) is specified, each group is assumed to have errors that follow the same AR(1) process; that is, the autocorrelation parameter is the same for all groups.

```
. xtgls invest market stock, i(company) t(time) panels(hetero) corr(ar1)

Cross-sectional time-series FGLS regression

Coefficients:  generalized least squares
Panels:        heteroskedastic
Correlation:   common AR(1) coefficient for all panels  (0.8651)

Estimated covariances      =        5          Number of obs      =         100
Estimated autocorrelations =        1          Number of groups   =           5
Estimated coefficients     =        3          Time periods       =          20
                                               Wald chi2(2)       =      119.69
Log likelihood             = -506.0909         Prob > chi2        =      0.0000
```

invest	Coef.	Std. Err.	z	P>|z|	[95% Conf. Interval]	
market	.0744315	.0097937	7.60	0.000	.0552362	.0936268
stock	.2874294	.0475391	6.05	0.000	.1942545	.3806043
_cons	-18.96238	17.64943	-1.07	0.283	-53.55464	15.62987

◁

(Continued on next page)

▷ Example

If corr(psar1) is specified, each group is assumed to have errors that follow a different AR(1) process.

```
. xtgls invest market stock, i(company) t(time) panels(iid) corr(psar1)

Cross-sectional time-series FGLS regression

Coefficients:  generalized least squares
Panels:        homoskedastic
Correlation:   panel-specific AR(1)
```

Estimated covariances =	1	Number of obs =	100
Estimated autocorrelations =	5	Number of groups =	5
Estimated coefficients =	3	Time periods =	20
		Wald chi2(2) =	252.93
Log likelihood = -543.1888		Prob > chi2 =	0.0000

invest	Coef.	Std. Err.	z	P>|z|	[95% Conf. Interval]
market	.0934343	.0097783	9.56	0.000	.0742693 .1125993
stock	.3838814	.0416775	9.21	0.000	.302195 .4655677
_cons	-10.1246	34.06675	-0.30	0.766	-76.8942 56.64499

◁

Saved Results

xtgls saves in e():

Scalars

e(N)	number of observations	e(df)	degrees of freedom
e(N_g)	number of groups	e(ll)	log likelihood
e(N_t)	number of time periods	e(g_max)	largest group size
e(N_miss)	number of missing observations	e(g_min)	smallest group size
e(n_cf)	number of estimated coefficients	e(g_avg)	average group size
e(n_cv)	number of estimated covariances	e(chi2)	χ^2
e(n_cr)	number of estimated correlations	e(df_pear)	degrees of freedom for Pearson χ^2

Macros

e(cmd)	xtgls	e(ivar)	variable denoting groups
e(depvar)	name of dependent variable	e(tvar)	variable denoting time
e(title)	title in estimation output	e(wtype)	weight type
e(coefftype)	estimation scheme	e(wexp)	weight expression
e(corr)	correlation structure	e(chi2type)	Wald; type of model χ^2 test
e(vt)	panel option	e(rho)	ρ
e(rhotype)	type of estimated correlation	e(predict)	program used to implement predict

Matrices

e(b)	coefficient vector	e(Sigma)	$\widehat{\Sigma}$ matrix
e(V)	variance–covariance matrix of the estimators		

Functions

e(sample)	marks estimation sample

Methods and Formulas

xtgls is implemented as an ado-file.

The GLS results are given by

$$\widehat{\beta}_{\text{GLS}} = (\mathbf{X}'\widehat{\Omega}^{-1}\mathbf{X})^{-1}\mathbf{X}'\widehat{\Omega}^{-1}\mathbf{y}$$
$$\widehat{\text{Var}}(\widehat{\beta}_{\text{GLS}}) = (\mathbf{X}'\widehat{\Omega}^{-1}\mathbf{X})^{-1}$$

For all our models, the Ω matrix may be written in terms of the Kronecker product:

$$\Omega = \boldsymbol{\Sigma}_{m\times m} \otimes \mathbf{I}_{T_i \times T_i}$$

The estimated variance matrix is then obtained by substituting the estimator $\widehat{\Sigma}$ for Σ, where

$$\widehat{\Sigma}_{i,j} = \frac{\widehat{\epsilon}_i{}'\widehat{\epsilon}_j}{T}$$

The residuals used in estimating Σ are first obtained from OLS regression. If the estimation is iterated, then residuals are obtained from the last fitted model.

Maximum likelihood estimates may be obtained by iterating the FGLS estimates to convergence (for models with no autocorrelation, corr(0)).

Note that the GLS estimates and their associated standard errors are calculated using $\widehat{\Sigma}^{-1}$. As Beck and Katz (1995) point out, the Σ matrix is of rank at most $\min(T, m)$ when you use the panels(correlated) option, so for the GLS results to be valid (not based on a generalized inverse), T must be at least as large as m (you need at least as many time period observations as there are panels).

Beck and Katz (1995) suggest using OLS parameter estimates with asymptotic standard errors that are correct for correlation between the panels. This estimation can be performed with the xtpcse command; see [XT] **xtpcse**.

References

Baum, C. F. 2001. Residual diagnostics. *The Stata Journal* 1: 101–104.

Beck, N. and J. N. Katz. 1995. What to do (and not to do) with time-series cross-section data. *American Political Science Review* 89: 634–647.

Davidson, R. and J. G. MacKinnon. 1993. *Estimation and Inference in Econometrics.* New York: Oxford University Press.

Greene, W. H. 2003. *Econometric Analysis.* 5th ed. Upper Saddle River, NJ: Prentice–Hall.

Grunfeld, Y. and Z. Griliches. 1960. Is aggregation necessarily bad? *Review of Economics and Statistics* 42: 1–13.

Judge, G. G., W. E. Griffiths, R. C. Hill, H. Lütkepohl, and T.-C. Lee. 1985. *The Theory and Practice of Econometrics.* 2d ed. New York: John Wiley & Sons.

Maddala, G. S. 1992. *Introduction to Econometrics.* 2d ed. New York: Macmillan.

Also See

Complementary:	[XT] **xtdata**, [XT] **xtdes**, [XT] **xtsum**, [XT] **xttab**,
	[R] **adjust**, [R] **lincom**, [R] **mfx**, [R] **nlcom**, [R] **predict**, [R] **predictnl**,
	[R] **test**, [R] **testnl**, [R] **vce**
Related:	[XT] **xtgee**, [XT] **xtpcse**, [XT] **xtreg**, [XT] **xtregar**,
	[R] **regress**,
	[SVY] **svy estimators**,
	[TS] **newey**, [TS] **prais**
Background:	[U] **16.5 Accessing coefficients and standard errors**,
	[U] **23 Estimation and post-estimation commands**,
	[XT] **xt**

Title

> **xthtaylor** — Hausman–Taylor estimator for error components models

Syntax

xthtaylor *depvar varlist* [*weight*] [if *exp*] [in *range*] , <u>en</u>dog(*varlist*_{endogenous})

$\qquad$ [[<u>const</u>ant(*varlist*_{invariant}) | <u>vary</u>ing(*varlist*_{varying})] <u>ama</u>curdy <u>small</u>

$\qquad$ <u>noc</u>onstant i(*varname*) t(*varname*) <u>le</u>vel(*#*)]

by ...: may be used with xthtaylor; see [R] **by**.

You must tsset your data before using xthtaylor in order to use time-series operators; see [TS] **tsset**.

depvar and all *varlist*s may contain time-series operators; see [U] **14.4.3 Time-series varlists**.

iweights and fweights are allowed unless the amacurdy option is specified; fweights must be constant within panel; see [U] **14.1.6 weight**.

xthtaylor shares the features of all estimation commands; see [U] **23 Estimation and post-estimation commands**.

Syntax for predict

predict [*type*] *newvarname* [if *exp*] [in *range*] [, *statistic*]

where *statistic* is

xb	$\mathbf{X}_{it}\widehat{\beta} + \mathbf{Z}_i\widehat{\delta}$, fitted values (the default)
stdp	standard error of the fitted values
ue	$\widehat{\mu}_i + \widehat{\epsilon}_{it}$, the combined residual
∗ xbu	$\mathbf{X}_{it}\widehat{\beta} + \mathbf{Z}_i\widehat{\delta} + \widehat{\mu}_i$, prediction including effect
∗ u	$\widehat{\mu}_i$, the random error component
∗ e	$\widehat{\epsilon}_{it}$, prediction of the idiosyncratic error component

Unstarred statistics are available both in and out of sample; type predict ... if e(sample) ... if wanted only for the estimation sample. Starred statistics are calculated only for the estimation sample even when if e(sample) is not specified.

Description

xthtaylor fits panel-data random-effects models in which some of the covariates are correlated with the unobserved individual-level random effect. The estimators, originally proposed by Hausman and Taylor (1981) and by Amemiya and MaCurdy (1986), are based on instrumental variables. By default, xthtaylor uses the Hausman–Taylor estimator. When the amacurdy option is specified, xthtaylor uses the Amemiya–MaCurdy estimator.

Although the estimators implemented in xthtaylor and the estimators implemented in xtivreg (see [XT] **xtivreg**), both use the method of instrumental variables. xthtaylor and xtivreg are designed for very different problems. The estimators that are implemented in xtivreg assume that a subset of the explanatory variables in the model are correlated with the idiosyncratic error ϵ_{it}. In contrast, the Hausman–Taylor and Amemiya–MaCurdy estimators that are implemented in xthtaylor assume that a subset of the explanatory variables are correlated with the individual-level random-effects, u_i, but that none of the explanatory variables are correlated with the idiosyncratic error, ϵ_{it}.

Options

endog($varlist_{\text{endogenous}}$) specifies the subset of explanatory variables in *varlist* that are to be treated as endogenous variables; i.e., the explanatory variables that are assumed to be correlated with the unobserved random effect. endog($varlist_{\text{endogenous}}$) is a required option.

constant($varlist_{\text{invariant}}$) specifies the subset of variables in *varlist* that are time invariant; that is, constant within panel. By using this option, you are not only asserting that the variables specified in $varlist_{\text{invariant}}$ are time invariant, but also that all other variables in *varlist* are time varying. If this assertion proves to be false, xthtaylor will refuse to perform the estimation and will issue an error message. xthtaylor automatically detects which variables are time invariant and which are not. However, there are times when users might want to check their understanding of the data and specify which variables are time invariant and which are not.

varying($varlist_{\text{varying}}$) specifies the subset of variables in *varlist* that are time varying. By using this option, you are not only asserting that the variables specified in $varlist_{\text{varying}}$ are time varying, but also that all other variables in *varlist* are time invariant. If this assertion proves to be false, xthtaylor will refuse to perform the estimation and will issue an error message. xthtaylor automatically detects which variables are time varying and which are not. However, there are times when users might want to check their understanding of the data and specify which variables are time varying and which are not.

amacurdy specifies that the Amemiya–MaCurdy estimator is to be used. This estimator uses extra instruments to gain efficiency at the cost of additional assumptions on the data-generating process. This option may only be specified for samples containing balanced panels, and weights may not be specified. The panels must also have a common initial time period.

small specifies that the p-values from the Wald tests in the output and all subsequent Wald tests obtained via test should use t and F distributions instead of the large sample normal and χ^2 distributions. By default, the p-values are obtained using the normal and χ^2 distributions.

noconstant suppresses the constant term (intercept) in the model.

i(*varname*) specifies the variable name that contains the unit to which the observation belongs. You can specify the i() option the first time you estimate, or you can use the iis command to set i() beforehand. Note that it is not necessary to specify i() if the data have been previously tsset, or if iis has been previously specified—in these cases, the group variable is taken from the previous setting. See [XT] **xt**.

t(*varname*) specifies the variable that contains the time at which the observation was made. This option is not necessary for the Hausman–Taylor estimator, but it is required for the Amemiya–MaCurdy estimator. You can specify the t() option the first time you estimate or you can use the tis command to set t() beforehand. Alternatively, you may tsset the data prior to estimation (see [TS] **tsset**).

level(#) specifies the confidence level, in percent, for confidence intervals of the coefficients. The default is level(95) or as set by set level; see [U] **23.6 Specifying the width of confidence intervals**.

Options for predict

xb, the default, calculates the linear prediction; that is, $\mathbf{X}_{it}\widehat{\beta} + \mathbf{Z}_{it}\widehat{\delta}$.

stdp calculates the standard error of the linear prediction.

ue calculates the prediction of $\widehat{\mu}_i + \widehat{\epsilon}_{it}$.

xbu calculates the prediction of $\mathbf{X}_{it}\widehat{\boldsymbol{\beta}} + \mathbf{Z}_{it}\widehat{\boldsymbol{\delta}} + \widehat{\nu}_i$, the prediction including the random effect.

u calculates the prediction of $\widehat{\mu}_i$, the estimated random effect.

e calculates the prediction of $\widehat{\epsilon}_{it}$.

Remarks

If you have not read [XT] **xt**, please do so.

Consider a random-effects model of the form

$$y_{it} = \mathbf{X}_{1it}\boldsymbol{\beta}_1 + \mathbf{X}_{2it}\boldsymbol{\beta}_2 + \mathbf{Z}_{1i}\boldsymbol{\delta}_1 + \mathbf{Z}_{2i}\boldsymbol{\delta}_2 + \mu_i + \epsilon_{it} \tag{1}$$

where

$\mathbf{X}_{1it}$ is a $1 \times k_1$ vector of observations on exogenous, time-varying variables assumed to be uncorrelated with μ_i and ϵ_{it};

$\mathbf{X}_{2it}$ is a $1 \times k_2$ vector of observations on endogenous, time-varying variables assumed to be (possibly) correlated with μ_i but orthogonal to ϵ_{it};

$\mathbf{Z}_{1i}$ is a $1 \times g_1$ vector of observations on exogenous, time-invariant variables assumed to be uncorrelated with μ_i and ϵ_{it};

$\mathbf{Z}_{2i}$ is a $1 \times g_2$ vector of observations on endogenous, time-invariant variables assumed to be (possibly) correlated μ_i but orthogonal to ϵ_{it};

μ_i is the unobserved, panel-level random effect that is assumed to have zero mean, finite variance σ_μ^2, and to be independently and identically distributed (i.i.d.) over the panels;

ϵ_{it} is the idiosyncratic error that is assumed to have zero mean, finite variance σ_ϵ^2, and to be i.i.d. over all the observations in the data;

$\boldsymbol{\beta}_1, \boldsymbol{\beta}_2, \boldsymbol{\delta}_1$, and $\boldsymbol{\delta}_2$ are $k_1 \times 1$, $k_2 \times 1$, $g_1 \times 1$, and $g_2 \times 1$ coefficient vectors, respectively; and

$i = 1, \ldots, n$, where n is the number of panels in the sample and, for each i, $t = 1, \ldots, T_i$.

Since $\mathbf{X}_{2it}$ and $\mathbf{Z}_{2i}$ are possibly correlated with μ_i, the simple random-effects estimators, xtreg, re and xtreg, mle, are generally not consistent for the parameters in this model. Since the within estimator, xtreg, fe, removes the μ_i by mean-differencing the data before estimating $\boldsymbol{\beta}_1$ and $\boldsymbol{\beta}_2$, it is consistent for these parameters. However, in the process of removing the μ_i, the within estimator also eliminates the $\mathbf{Z}_{1i}$ and the $\mathbf{Z}_{2i}$. Thus, it cannot estimate $\boldsymbol{\delta}_1$ nor $\boldsymbol{\delta}_2$. The Hausman–Taylor and Amemiya–MaCurdy estimators implemented in xthtaylor are designed to resolve just this problem.

As mentioned previously, the within estimator consistently estimates $\boldsymbol{\beta}_1$ and $\boldsymbol{\beta}_2$. Using these estimates, we can obtain the within-residuals, called $\widehat{d}_i$. Intermediate, albeit consistent, estimates of $\boldsymbol{\delta}_1$ and $\boldsymbol{\delta}_2$, called $\widehat{\boldsymbol{\delta}}_{1\mathrm{IV}}$ and $\widehat{\boldsymbol{\delta}}_{2\mathrm{IV}}$, respectively, are obtained by regressing the within-residuals on $\mathbf{Z}_{1i}$ and $\mathbf{Z}_{2i}$, using $\mathbf{X}_{1it}$ and $\mathbf{Z}_{1i}$ as instruments. The order condition for identification requires that the number of variables in $\mathbf{X}_{1it}$, k_1, be at least as large as the number of elements in $\mathbf{Z}_{2i}$, g_2, and there needs to be sufficient correlation between the instruments and $\mathbf{Z}_{2i}$ to avoid a weak instrument problem.

The within estimates of $\boldsymbol{\beta}_1$ and $\boldsymbol{\beta}_2$ and the intermediate estimates $\widehat{\boldsymbol{\delta}}_{1\mathrm{IV}}$ and $\widehat{\boldsymbol{\delta}}_{2\mathrm{IV}}$ can be used to obtain sets of within and overall residuals. These two sets of residuals can be used to estimate the variance components (see *Methods and Formulas* for details).

The estimated variance components can then be used to perform a GLS transform on each of the variables. For what follows, define the general notation $\breve{w}_{it}$ to represent the GLS transform of the variable w_{it}, $\overline{w}_i$ to represent the within-panel mean of w_{it}, and $\widetilde{w}_{it}$ to represent the within transform of w_{it}. Using this notational convention, the Hausman–Taylor (1981) estimator of the coefficients of interest can be obtained by the instrumental variables regression

$$\breve{y}_{it} = \breve{\mathbf{X}}_{1it}\boldsymbol{\beta}_1 + \breve{\mathbf{X}}_{2it}\boldsymbol{\beta}_2 + \breve{\mathbf{Z}}_{1i}\boldsymbol{\delta}_1 + \breve{\mathbf{Z}}_{2i}\boldsymbol{\delta}_2 + \breve{\mu}_i + \breve{\epsilon}_{it} \tag{2}$$

using $\widetilde{\mathbf{X}}_{1it}$, $\widetilde{\mathbf{X}}_{2it}$, $\overline{\mathbf{X}}_{1i}$, $\overline{\mathbf{X}}_{2i}$, and $\mathbf{Z}_{1i}$ as instruments.

In order for the instruments to be valid, this estimator requires that $\overline{\mathbf{X}}_{1i.}$ and $\mathbf{Z}_{1i}$ be uncorrelated with the random-effect μ_i. More precisely, the instruments are valid when

$$\text{plim}_{n\to\infty} \frac{1}{n} \sum_{i=1}^{n} \overline{\mathbf{X}}_{1i.}\mu_i = 0$$

and

$$\text{plim}_{n\to\infty} \frac{1}{n} \sum_{i=1}^{n} \mathbf{Z}_{1i}\mu_i = 0$$

Amemiya and MaCurdy (1986) placed stricter requirements on the instruments that vary within panels, and obtained a more efficient estimator. Specifically, Amemiya and MaCurdy (1986) assumed that $\mathbf{X}_{1it}$ is orthogonal to μ_i in every time period; i.e., $\text{plim}_{n\to\infty} \frac{1}{n} \sum_{i=1}^{n} \mathbf{X}_{1it}\mu_i = 0$ for $t = 1,\ldots,T$. With this restriction, they derived the Amemiya–MaCurdy estimator as the instrumental variables regression of (2) using instruments $\widetilde{\mathbf{X}}_{1it}$, $\widetilde{\mathbf{X}}_{2it}$, $\mathbf{X}_{1it}^*$, and $\mathbf{Z}_{1i}$. Note that the order condition for the Amemiya–MaCurdy estimator is now $Tk_1 > g_2$. xthtaylor will use the Amemiya–MaCurdy estimator when the amacurdy option is specified.

▷ Example

This example replicates the results of Baltagi and Khanti-Akom (1990, Table II, column HT) using 595 observations on individuals over the period 1976–1982 that were extracted from the Panel Study of Income Dynamics (PSID). In the model, the log-transformed wage lwage is assumed to be a function of how long the person has worked for a firm, wks; binary variables indicating whether a person lives in a large metropolitan area or in the south, smsa and south; marital status, ms; years of education, ed; a quadratic of work experience, exp and exp2; occupation, occ; a binary variable indicating employment in a manufacture industry, ind; a binary variable indicating that wages are set by a union contract, union; a binary variable indicating gender, fem; and a binary variable indicating whether the individual is African-American, blk.

It is suspected that the time-varying variables exp, exp2, wks, ms, and union are all correlated with the unobserved individual random effect. We can inspect these variables to see if they exhibit sufficient within-panel variation to serve as their own instruments.

(Continued on next page)

```
. use http://www.stata-press.com/data/r8/psidextract
. xtsum exp exp2 wks ms union, i(id)
```

Variable		Mean	Std. Dev.	Min	Max	Observations	
exp	overall	19.85378	10.96637	1	51	N =	4165
	between		10.79018	4	48	n =	595
	within		2.00024	16.85378	22.85378	T =	7
exp2	overall	514.405	496.9962	1	2601	N =	4165
	between		489.0495	20	2308	n =	595
	within		90.44581	231.405	807.405	T =	7
wks	overall	46.81152	5.129098	5	52	N =	4165
	between		3.284016	31.57143	51.57143	n =	595
	within		3.941881	12.2401	63.66867	T =	7
ms	overall	.8144058	.3888256	0	1	N =	4165
	between		.3686109	0	1	n =	595
	within		.1245274	-.0427371	1.671549	T =	7
union	overall	.3639856	.4812023	0	1	N =	4165
	between		.4543848	0	1	n =	595
	within		.1593351	-.4931573	1.221128	T =	7

We are also going to assume that the exogenous variables occ, south, smsa, ind, fem, and blk are instruments for the endogenous, time-invariant variable ed. The output below indicates that while fem appears to be a weak instrument, the remaining instruments are probably sufficiently correlated to identify the coefficient on ed. (See Baltagi and Khanti-Akom (1990) for further discussion.)

```
. correlate fem blk occ south smsa ind ed
(obs=4165)
```

	fem	blk	occ	south	smsa	ind	ed
fem	1.0000						
blk	0.2086	1.0000					
occ	-0.0847	0.0837	1.0000				
south	0.0516	0.1218	0.0413	1.0000			
smsa	0.1044	0.1154	-0.2018	-0.1350	1.0000		
ind	-0.1778	-0.0475	0.2260	-0.0769	-0.0689	1.0000	
ed	-0.0012	-0.1196	-0.6194	-0.1216	0.1843	-0.2365	1.0000

For our purposes, we will assume that the correlations are strong enough and proceed with the estimation. The output below gives the Hausman–Taylor estimates for this model.

(Continued on next page)

```
. xthtaylor lwage occ south smsa ind exp exp2 wks ms union fem blk ed,
> endog(exp exp2 wks ms union ed)
```

| Hausman-Taylor estimation | | | | Number of obs | = | 4165 |
| Group variable (i): id | | | | Number of groups | = | 595 |

```
                                         Obs per group: min =          7
                                                        avg =          7
                                                        max =          7

Random effects u_i ~ i.i.d.              Wald chi2(12)       =    6891.87
                                         Prob > chi2         =     0.0000
```

lwage	Coef.	Std. Err.	z	P>|z|	[95% Conf. Interval]	
TVexogenous						
occ	-.0207047	.0137809	-1.50	0.133	-.0477149	.0063055
south	.0074398	.031955	0.23	0.816	-.0551908	.0700705
smsa	-.0418334	.0189581	-2.21	0.027	-.0789906	-.0046761
ind	.0136039	.0152374	0.89	0.372	-.0162608	.0434686
TVendogenous						
exp	.1131328	.002471	45.79	0.000	.1082898	.1179758
exp2	-.0004189	.0000546	-7.67	0.000	-.0005259	-.0003119
wks	.0008374	.0005997	1.40	0.163	-.0003381	.0020129
ms	-.0298508	.01898	-1.57	0.116	-.0670508	.0073493
union	.0327714	.0149084	2.20	0.028	.0035514	.0619914
TIexogenous						
fem	-.1309236	.126659	-1.03	0.301	-.3791707	.1173234
blk	-.2857479	.1557019	-1.84	0.066	-.5909179	.0194221
TIendogenous						
ed	.137944	.0212485	6.49	0.000	.0962977	.1795902
_cons	2.912726	.2836522	10.27	0.000	2.356778	3.468674
sigma_u	.94180304					
sigma_e	.15180273					
rho	.97467788	(fraction of variance due to u_i)				

```
note:  TV refers to time-varying; TI refers to time-invariant.
```

The estimated σ_μ and σ_ϵ are 0.9418 and 0.1518, respectively. This indicates that a very large fraction of the total error variance is attributed to μ_i. The z statistics indicate that a number of the coefficients may not be significantly different from zero. It is interesting to note that while the coefficients on the time-invariant variables fem and blk have relatively large standard errors, the standard error for the coefficient on ed is relatively small.

Baltagi and Khanti-Akom (1990) also presented evidence that the efficiency gains of the Amemiya–MaCurdy estimator over the Hausman–Taylor estimator are small for these data. This point is especially important given the additional restrictions that the estimator places on the data-generating process. The output below replicates the Baltagi and Khanti-Akom (1990) results from column AM of Table II.

(Continued on next page)

```
. set matsize 100

. xthtaylor lwage occ south smsa ind exp exp2 wks ms union fem blk ed,
> endog(exp exp2 wks ms union ed) amacurdy t(t)
```

| Amemiya-MaCurdy estimation | | | | Number of obs | | = | 4165 |
| Group variable (i): id | | | | Number of groups | | = | 595 |

		Obs per group: min =	7
		avg =	7
		max =	7

| Random effects u_i ~ i.i.d. | | | | Wald chi2(12) | | = | 6879.20 |
| | | | | Prob > chi2 | | = | 0.0000 |

lwage	Coef.	Std. Err.	z	P>|z|	[95% Conf. Interval]	
TVexogenous						
occ	-.0208498	.0137653	-1.51	0.130	-.0478292	.0061297
south	.0072818	.0319365	0.23	0.820	-.0553126	.0698761
smsa	-.0419507	.0189471	-2.21	0.027	-.0790864	-.0048149
ind	.0136289	.015229	0.89	0.371	-.0162194	.0434771
TVendogenous						
exp	.1129704	.0024688	45.76	0.000	.1081316	.1178093
exp2	-.0004214	.0000546	-7.72	0.000	-.0005283	-.0003145
wks	.0008381	.0005995	1.40	0.162	-.0003368	.002013
ms	-.0300894	.0189674	-1.59	0.113	-.0672649	.0070861
union	.0324752	.0148939	2.18	0.029	.0032837	.0616667
TIexogenous						
fem	-.132008	.1266039	-1.04	0.297	-.380147	.1161311
blk	-.2859004	.1554857	-1.84	0.066	-.5906468	.0188459
TIendogenous						
ed	.1372049	.0205695	6.67	0.000	.0968894	.1775205
_cons	2.927338	.2751274	10.64	0.000	2.388098	3.466578
sigma_u	.94180304					
sigma_e	.15180273					
rho	.97467788	(fraction of variance due to u_i)				

note: TV refers to time-varying; TI refers to time-invariant.

◁

❑ Technical Note

It was mentioned earlier that insufficient correlation between an endogenous variable and the instrument(s) can give rise to a weak instrument problem. For example, suppose we simulate data for a model of the form

$$y = 3 + 3x_{1a} + 3x_{1b} + 3x_2 + 3z_1 + 3z_2 + u_i + e_{it}$$

and purposely construct the instruments so that they exhibit very little correlation with the endogenous variable z_2.

(Continued on next page)

```
. use http://www.stata-press.com/data/r8/xthtaylor1
. correlate ui z1 z2 x1a x1b x2 eit
(obs=10000)
```

	ui	z1	z2	x1a	x1b	x2	eit
ui	1.0000						
z1	0.0268	1.0000					
z2	0.8777	0.0286	1.0000				
x1a	-0.0145	0.0065	-0.0034	1.0000			
x1b	0.0026	0.0079	0.0038	-0.0030	1.0000		
x2	0.8765	0.0191	0.7671	-0.0192	0.0037	1.0000	
eit	0.0060	-0.0198	0.0123	-0.0100	-0.0138	0.0092	1.0000

In the output below, you can see that weak instruments have serious consequences on the estimates produced by xthtaylor. Note that the estimate of the coefficient on z2 is 3 times larger than its true value and that its standard error is rather large. Without sufficient correlation between the endogenous variable and its instruments in a given sample, there is insufficient information with which to identify the parameter. Also, given the results of Stock, et al. (2002), weak instruments will cause serious size distortion in any tests performed.

```
. xthtaylor yit x1a x1b x2 z1 z2, endog(x2 z2) i(id)
```

Hausman-Taylor estimation		Number of obs	=	10000
Group variable (i): id		Number of groups	=	1000
		Obs per group: min =		10
		avg =		10
		max =		10
Random effects u_i ~ i.i.d.		Wald chi2(5)	=	24172.91
		Prob > chi2	=	0.0000

yit	Coef.	Std. Err.	z	P>|z|	[95% Conf. Interval]	
TVexogenous						
x1a	2.959736	.0330233	89.63	0.000	2.895011	3.02446
x1b	2.953891	.0333051	88.69	0.000	2.888614	3.019168
TVendogenous						
x2	3.022685	.033085	91.36	0.000	2.957839	3.08753
TIexogenous						
z1	2.709179	.587031	4.62	0.000	1.55862	3.859739
TIendogenous						
z2	9.525973	8.572966	1.11	0.266	-7.276732	26.32868
_cons	2.837072	.4276595	6.63	0.000	1.998875	3.675269
sigma_u	8.729479					
sigma_e	3.1657492					
rho	.88377062	(fraction of variance due to u_i)				

```
note:  TV refers to time-varying; TI refers to time-invariant.
```

❑

▷ Example

Now, let's consider why we might want to specify the constant($varlist_{\text{invariant}}$) option. For this example, we will use simulated data. In the output below, we fit a model over the full sample. Note, in particular, the placement in the output of the coefficient on the exogenous variable x1c.

```
. use http://www.stata-press.com/data/r8/xthtaylor2
. xthtaylor yit x1a x1b x1c x2 z1 z2, endog(x2 z2) i(id)
```

Hausman-Taylor estimation   Number of obs   =   10000
Group variable (i): id   Number of groups   =   1000

                                       Obs per group: min =   10
                                                avg =   10
                                                max =   10
Random effects u_i ~ i.i.d.   Wald chi2(6)   =   10341.63
                                        Prob > chi2   =   0.0000

yit	Coef.	Std. Err.	z	P>|z|	[95% Conf. Interval]	
TVexogenous						
x1a	3.023647	.0570274	53.02	0.000	2.911875	3.135418
x1b	2.966666	.0572659	51.81	0.000	2.854427	3.078905
x1c	.2355318	.123502	1.91	0.057	-.0065276	.4775912
TVendogenous						
x2	14.17476	3.128385	4.53	0.000	8.043234	20.30628
TIexogenous						
z1	1.741709	.4280022	4.07	0.000	.9028398	2.580578
TIendogenous						
z2	7.983849	.6970903	11.45	0.000	6.617577	9.350121
_cons	2.146038	.3794179	5.66	0.000	1.402393	2.889684
sigma_u	5.6787791					
sigma_e	3.1806188					
rho	.76120931	(fraction of variance due to u_i)				

note: TV refers to time-varying; TI refers to time-invariant.

Now, suppose that we want to fit the model using only the first eight periods. Notice in the output below that x1c now appears under the TIexogenous heading rather than the TVexogenous heading, because x1c is time invariant in the subsample defined by t<9.

(Continued on next page)

```
. xthtaylor yit x1a x1b x1c x2 z1 z2 if t<9, endog(x2 z2) i(id)
```

| Hausman-Taylor estimation | | Number of obs | = | 8000 |
| Group variable (i): id | | Number of groups | = | 1000 |

```
                                      Obs per group: min =          8
                                                     avg =          8
                                                     max =          8
```

| Random effects u_i ~ i.i.d. | | Wald chi2(6) | = | 15354.87 |
| | | Prob > chi2 | = | 0.0000 |

yit	Coef.	Std. Err.	z	P>|z|	[95% Conf. Interval]	
TVexogenous						
x1a	3.051966	.0367026	83.15	0.000	2.98003	3.123901
x1b	2.967822	.0368144	80.62	0.000	2.895667	3.039977
TVendogenous						
x2	.7361217	3.199764	0.23	0.818	-5.5353	7.007543
TIexogenous						
x1c	3.215907	.5657191	5.68	0.000	2.107118	4.324696
z1	3.347644	.5819756	5.75	0.000	2.206992	4.488295
TIendogenous						
z2	2.010578	1.143982	1.76	0.079	-.231586	4.252742
_cons	3.257004	.5295828	6.15	0.000	2.219041	4.294967
sigma_u	15.445594					
sigma_e	3.175083					
rho	.95945606	(fraction of variance due to u_i)				

```
note:  TV refers to time-varying; TI refers to time-invariant.
```

To prevent a variable from becoming time invariant without notice, one can use either the constant(*varlist*$_\text{invariant}$) option or the varying(*varlist*$_\text{varying}$) option. Recall that constant(*varlist*$_\text{invariant}$) specifies the subset of variables in *varlist* that are time invariant, and requires the remaining variables in *varlist* to be time varying. If you specify the constant(*varlist*$_\text{invariant}$) option and any of the variables contained in *varlist*$_\text{invariant}$ are, in fact, time varying, or if any of the variables not contained in *varlist*$_\text{invariant}$ are, in fact, time invariant, xthtaylor will refuse to perform the estimation and will issue an error message.

```
. xthtaylor yit x1a x1b x1c x2 z1 z2 if t<9, endog(x2 z2) i(id) constant(z1 z2)
x1c not included in -constant()-.
```

Analogous behavior occurs when using the varying(*varlist*$_\text{varying}$) option.

◁

(Continued on next page)

Saved Results

xthtaylor saves in e():

Scalars

e(N)	number of observations	e(chi2)	χ^2
e(N_g)	number of groups	e(rho)	ρ
e(df_m)	model degrees of freedom	e(sigma_u)	panel-level standard deviation
e(def_r)	residual degrees of freedom	e(sigma_e)	standard deviation of ϵ_{it}
e(g_max)	largest group size	e(F)	model F (small only)
e(g_min)	smallest group size	e(Tbar)	harmonic mean of group sizes
e(g_avg)	average group size		

Macros

e(cmd)	xthtaylor	e(tvar)	time variable, amacurdy only
e(depvar)	name of dependent variable	e(wtype)	weight type
e(title)	Hausman-Taylor	e(wexp)	weight expression
	or Amemiya-MaCurdy	e(chi2type)	Wald; type of model χ^2 test
e(TVexogenous)	exogenous time-varying variables	e(TVendogenous)	endogenous time-varying variables
e(TIexogenous)	exogenous time-invariant variables	e(TIendogenous)	endogenous time-invariant variables
e(ivar)	variable denoting groups	e(predict)	program used to implement predict

Matrices

e(b)	coefficient vector	e(V)	variance–covariance matrix of the estimators

Functions

e(sample)	marks estimation sample

Methods and Formulas

xthtaylor is implemented as an ado-file.

Consider an error components model of the form

$$y_{it} = \mathbf{X}_{1it}\boldsymbol{\beta}_1 + \mathbf{X}_{2it}\boldsymbol{\beta}_2 + \mathbf{Z}_{1i}\boldsymbol{\delta}_1 + \mathbf{Z}_{2i}\boldsymbol{\delta}_2 + \mu_i + \epsilon_{it} \tag{3}$$

for $i = 1, \ldots, n$ and, for each i, $t = 1, \ldots, T_i$, of which T_i periods are observed; n is the number of panels in the sample. The covariates in $\mathbf{X}$ are time varying, and the covariates in $\mathbf{Z}$ are time-invariant. Both $\mathbf{X}$ and $\mathbf{Z}$ are decomposed into two parts. The covariates in $\mathbf{X}_1$ and $\mathbf{Z}_1$ are assumed to be uncorrelated with μ_i and e_{it}, while the covariates in $\mathbf{X}_2$ and $\mathbf{Z}_2$ are allowed to be correlated with μ_i, but not with ϵ_{it}. Hausman and Taylor (1981) suggested an instrumental variable estimator for this model.

Recall that, for some variable w, the within transformation of w is defined as

$$\widetilde{w}_{it} = w_{it} - \overline{w}_{i.} \qquad \overline{w}_{i.} = n^{-1} \sum_{t=1}^{T_i} w_{it}$$

Since the within estimator removes $\mathbf{Z}$, the within transformation reduces the model to

$$\widetilde{y}_{it} = \widetilde{\mathbf{X}}_{1it}\boldsymbol{\beta}_1 + \widetilde{\mathbf{X}}_{2it}\boldsymbol{\beta}_2 + \widetilde{\epsilon}_{it}$$

The within estimators $\widehat{\boldsymbol{\beta}}_{1w}$ and $\widehat{\boldsymbol{\beta}}_{2w}$ are consistent for $\boldsymbol{\beta}_1$ and $\boldsymbol{\beta}_2$, but they may not be efficient. Also, note that the within estimator cannot estimate $\boldsymbol{\delta}_1$ and $\boldsymbol{\delta}_2$.

The within estimator can be used to obtain the within residuals

$$\widetilde{d}_{it} = \widetilde{y}_{it} - \widetilde{\mathbf{X}}_{1it}\widehat{\boldsymbol{\beta}}_{1w} - \widetilde{\mathbf{X}}_{2it}\widehat{\boldsymbol{\beta}}_{2w}$$

which allows one to estimate the variance of the idiosyncratic error component, σ_ϵ^2, as

$$\widehat{\sigma}_\epsilon^2 = \frac{RSS}{N-n}$$

where RSS is the residual sum of squares from the within regression and N is the total number of observations in the sample.

Regressing $\widetilde{d}_{it}$ on $\mathbf{Z}_1$ and $\mathbf{Z}_2$, using $\mathbf{X}_1$ and $\mathbf{Z}_1$ as instruments, provides intermediate, consistent estimates of $\boldsymbol{\delta}_1$ and $\boldsymbol{\delta}_2$ that we shall call $\widehat{\boldsymbol{\delta}}_{1IV}$ and $\widehat{\boldsymbol{\delta}}_{2IV}$.

Using the within estimates, $\widehat{\boldsymbol{\delta}}_{1IV}$, and $\widehat{\boldsymbol{\delta}}_{2IV}$, one can obtain an estimate of the variance of the random effect, σ_μ^2. First, let

$$\widehat{e}_{it} = \left(y_{it} - \mathbf{X}_{1it}\widehat{\boldsymbol{\beta}}_{1w} - \mathbf{X}_{2it}\widehat{\boldsymbol{\beta}}_{2w} - \mathbf{Z}_{1it}\widehat{\boldsymbol{\delta}}_{1IV} - \mathbf{Z}_{2it}\widehat{\boldsymbol{\delta}}_{2IV} \right)$$

Then, define

$$s^2 = \frac{1}{N}\sum_{i=1}^{n}\sum_{t=1}^{T_i}\left(\frac{1}{T_i}\sum_{t=1}^{T_i}\widehat{e}_i \right)^2$$

Hausman and Taylor (1981) showed that in the case of balanced panels

$$\mathrm{plim}_{n\to\infty}s^2 = T\sigma_\mu^2 + \sigma_\epsilon^2$$

In the case of unbalanced panels,

$$\mathrm{plim}_{n\to\infty}s^2 = \overline{T}\sigma_\mu^2 + \sigma_\epsilon^2$$

where

$$\overline{T} = \frac{n}{\sum_{i=1}^{n}\frac{1}{T_i}}$$

After plugging in $\widehat{\sigma}_\epsilon^2$, our consistent estimate for σ_ϵ^2, a little algebra suggests the estimate

$$\widehat{\sigma}_\mu^2 = (s^2 - \widehat{\sigma}_\epsilon^2)(\overline{T})^{-1}$$

Define $\widehat{\theta}_i$ as

$$\widehat{\theta}_i = 1 - \left(\frac{\widehat{\sigma}_\epsilon^2}{\widehat{\sigma}_\epsilon^2 + T_i \widehat{\sigma}_\mu^2} \right)^{\frac{1}{2}}$$

With $\widehat{\theta}_i$ in hand, we can perform the standard random-effects GLS transform on each of the variables. The transform is given by

$$w_{it}^* = w_{it} - \widehat{\theta}_i \overline{w}_{i.}$$

where $\overline{w}_{i.}$ is the within-panel mean.

The Hausman–Taylor estimates of the coefficients in (3) are then obtained by fitting an instrumental variables regression of the GLS-transformed y_{it}^* on $\mathbf{X}_{it}^*$ and $\mathbf{Z}_{it}^*$, with instruments $\widetilde{\mathbf{X}}_{it}$, $\overline{\mathbf{X}}_{1i.}$, and $\mathbf{Z}_{1i}$.

The Amemiya–MaCurdy estimates of the coefficients in (3) are then obtained via an instrumental variables regression of the GLS-transformed y_{it}^* on $\mathbf{X}_{it}^*$ and $\mathbf{Z}_{it}^*$, using $\widetilde{\mathbf{X}}_{it}$, $\widecheck{\mathbf{X}}_{1it}$, and $\mathbf{Z}_{1i}$ as instruments, where $\widecheck{\mathbf{X}}_{1it} = \mathbf{X}_{1i1}, \mathbf{X}_{1i2}, \ldots, \mathbf{X}_{1iT_i}$. The order condition for the Amemiya–MaCurdy estimator is $Tk_1 > g_2$, and that this estimator is only available for balanced panels.

References

Amemiya, T. and T. MaCurdy. 1986. Instrumental-variable estimation of an error-components model. *Econometrica* 54(4): 869–880.

Baltagi, B. H. 2001. *Econometric Analysis of Panel Data*. 2d ed. New York: John Wiley & Sons.

Baltagi, B. H. and S. Khanti-Akom. 1990 On efficient estimation with panel data: an empirical comparison of instrumental variables estimators. *Journal of applied econometrics* 5: 401–406.

Hausman, J. A. and W. E. Taylor. 1981 Panel data and unobservable individual effects. *Econometrica* 49: 1377–1398.

Stock, J. H., J. H. Wright, and M. Yogo. 2002 A survey of weak instruments and weak identification in generalized method of moments. *Journal of Business & Economic Statistics* 20(4): 518–529.

Also See

Complementary:	[XT] **xtdata**, [XT] **xtdes**, [XT] **xtsum**, [XT] **xttab**,
	[R] **adjust**, [R] **lincom**, [R] **mfx**, [R] **nlcom**, [R] **predict**, [R] **predictnl**,
	[R] **test**, [R] **testnl**, [R] **vce**,
	[TS] **tsset**
Related:	[XT] **xtabond**, [XT] **xtgee**, [XT] **xtintreg**, [XT] **xtivreg**,
	[XT] **xtreg**, [XT] **xtregar**, [XT] **xttobit**
Background:	[U] **16.5 Accessing coefficients and standard errors**,
	[U] **23 Estimation and post-estimation commands**,
	[XT] **xt**

Title

> **xtintreg** — Random-effects interval data regression models

Syntax

Random-effects model

> xtintreg *depvar*~lower~ *depvar*~upper~ [*varlist*] [*weight*] [if *exp*] [in *range*]
>
> [, i(*varname*) quad(*#*) noconstant noskip level(*#*) offset(*varname*)
>
> intreg nolog *maximize_options*]

by ... : may be used with xtintreg; see [R] **by**.

iweights are allowed; see [U] **14.1.6 weight**. Note that weights must be constant within panels.

xtintreg shares the features of all estimation commands; see [U] **23 Estimation and post-estimation commands**.

Syntax for predict

> predict [*type*] *newvarname* [if *exp*] [in *range*] [, *statistics* nooffset]

where *statistic* is

xb	linear prediction assuming $\nu_i = 0$, the default
pr0(*a*,*b*)	$\Pr(a < y < b)$ assuming $\nu_i = 0$
e0(*a*,*b*)	$E(y \mid a < y < b)$ assuming $\nu_i = 0$
ystar0(*a*,*b*)	$E(y^*)$, $y^* = \max\{a, \min(y_j, b)\}$ assuming $\nu_i = 0$
stdp	standard error of the linear prediction
stdf	standard error of the linear forecast

where *a* and *b* may be numbers or variables; *a* missing ($a \geq .$) means $-\infty$, and *b* missing ($b \geq .$) means $+\infty$; see [U] **15.2.1 Missing values**.

These statistics are available both in and out of sample; type predict ... if e(sample) ... if wanted only for the estimation sample.

Description

xtintreg fits random-effects interval regression models. There is no command for a conditional fixed-effects model, as there does not exist a sufficient statistic allowing the fixed effects to be conditioned out of the likelihood. Unconditional fixed-effects intreg models may be fitted with the intreg command, with indicator variables for the panels. The appropriate indicator variables can be generated using tabulate or xi. However, unconditional fixed-effects estimates are biased.

Note: xtintreg is slow since it is calculated by quadrature; see *Methods and Formulas*. Computation time is roughly proportional to the number of points used for the quadrature. The default is quad(12). Simulations indicate that increasing it does not appreciably change the estimates for the coefficients or their standard errors. See [XT] **quadchk**.

Options

i(*varname*) specifies the variable name that contains the unit to which the observation belongs. You can specify the i() option the first time you estimate, or you can use the iis command to set i() beforehand. Note that it is not necessary to specify i() if the data have been previously tsset, or if iis has been previously specified—in these cases, the group variable is taken from the previous setting. See [XT] **xt**.

quad(*#*) specifies the number of points to use in the quadrature approximation of the integral.

The default is quad(12). The number specified must be an integer between 4 and 30, and must be no greater than the number of observations.

noconstant suppresses the constant term (intercept) in the model.

noskip specifies that a full maximum-likelihood model with only a constant for the regression equation be fitted. This model is not displayed, but is used as the base model to compute a likelihood-ratio test for the model test statistic displayed in the estimation header. By default, the overall model test statistic is an asymptotically equivalent Wald test of all the parameters in the regression equation being zero (except the constant). For many models, this option can substantially increase estimation time.

level(*#*) specifies the confidence level, in percent, for confidence intervals. The default is level(95) or as set by set level; see [U] **23.6 Specifying the width of confidence intervals**.

offset(*varname*) specifies that *varname* is to be included in the model with its coefficient constrained to be 1.

intreg specifies that a likelihood-ratio test comparing the random-effects model with the pooled (intreg) model should be included in the output.

nolog suppresses the iteration log.

maximize_options control the maximization process; see [R] **maximize**. Use the trace option to view parameter convergence. Use the ltol(*#*) option to relax the convergence criterion; the default is 1e−6 during specification searches.

Options for predict

xb, the default, calculates the linear prediction.

pr0(*a*,*b*) calculates estimates of $\Pr(a < y < b|\mathbf{x} = \mathbf{x}_{it}, \nu_i = 0)$, the probability that y would be observed in the interval (a, b), given the current values of the predictors, $\mathbf{x}_{it}$, and given a zero random effect; see *Remarks*. In the discussion that follows, these two conditions are taken to be implicit.

a and *b* may be specified as numbers or variable names; *lb* and *ub* are variable names;
pr0(20,30) calculates $\Pr(20 < y < 30)$;
pr0(*lb*,*ub*) calculates $\Pr(lb < y < ub)$; and
pr0(20,*ub*) calculates $\Pr(20 < y < ub)$.

a missing (*a* ≥ .) means $-\infty$; pr0(.,30) calculates $\Pr(-\infty < y < 30)$;
pr0(*lb*,30) calculates $\Pr(-\infty < y < 30)$ in observations for which *lb* ≥ .
(and calculates $\Pr(lb < y < 30)$ elsewhere).

b missing (*b* ≥ .) means $+\infty$; pr0(20,.) calculates $\Pr(+\infty > y > 20)$;
pr0(20,*ub*) calculates $\Pr(+\infty > y > 20)$ in observations for which *ub* ≥ .
(and calculates $\Pr(20 < y < ub)$ elsewhere).

e0(a,b) calculates estimates of $E(y \mid a < y < b, \mathbf{x} = \mathbf{x}_{it}, \nu_i = 0)$, the expected value of y conditional on y being in the interval (a, b), which is to say, y is censored. a and b are specified as they are for pr0().

ystar0(a,b) calculates estimates of $E(y^* \mid \mathbf{x} = \mathbf{x}_{it}, \nu_i = 0)$, where $y^* = a$ if $y \leq a$, $y^* = b$ if $y \geq b$, and $y^* = y$ otherwise, which is to say, y^* is the truncated version of y. a and b are specified as they are for pr0().

stdp calculates the standard error of the prediction. It can be thought of as the standard error of the predicted expected value or mean for the observation's covariate pattern. This is also referred to as the standard error of the fitted value.

stdf calculates the standard error of the forecast. This is the standard error of the point prediction for a single observation. It is commonly referred to as the standard error of the future or forecast value. By construction, the standard errors produced by stdf are always larger than those by stdp; see [R] **regress** *Methods and Formulas*.

nooffset is relevant only if you specified offset(*varname*) for xtintreg. It modifies the calculations made by predict so that they ignore the offset variable; the linear prediction is treated as $\mathbf{x}_{it}\boldsymbol{\beta}$ rather than $\mathbf{x}_{it}\boldsymbol{\beta} + \text{offset}_{it}$.

Remarks

Consider the linear regression model with panel-level random effects

$$y_{it} = \mathbf{x}_{it}\boldsymbol{\beta} + \nu_i + \epsilon_{it}$$

for $i = 1, \ldots, n$, $t = 1, \ldots, n_i$. The random effects, ν_i, are iid $N(0, \sigma_\nu^2)$, and ϵ_{it} are iid $N(0, \sigma_\epsilon^2)$ independently of ν_i. The observed data consist of the couples, (y_{1it}, y_{2it}), such that all that is known is that $y_{1it} \leq y_{it} \leq y_{2it}$, where y_{1it} is possibly $-\infty$ and y_{2it} is possibly $+\infty$.

▷ Example

We begin with the dataset nlswork described in [XT] **xt** and create two fictional dependent variables, where the wages are instead reported sometimes as ranges. The wages have been adjusted by a GNP deflator such that they are in terms of 1988 dollars, and have further been recoded such that some of the observations are known exactly, some are left-censored, some are right-censored, and some are known only in an interval.

We wish to fit a random-effects interval regression model of adjusted (log) wages:

```
. use http://www.stata-press.com/data/r8/nlswork3
(National Longitudinal Survey.  Young Women 14-26 years of age in 1968)
```

(Continued on next page)

```
. xtintreg ln_wage1 ln_wage2 union age grade not_smsa south southXt occ_code,
> i(id) noskip intreg nolog
```

Random-effects interval regression   Number of obs   =   19095
Group variable (i): idcode   Number of groups   =   4139

Random effects u_i ~ Gaussian   Obs per group: min =   1
                                                                  avg =   4.6
                                                                  max =   12

                                                   LR chi2(7)   =   3549.46
Log likelihood = -14856.934   Prob > chi2   =   0.0000

	Coef.	Std. Err.	z	P>|z|	[95% Conf.	Interval]
union	.1409746	.0068364	20.62	0.000	.1275755	.1543737
age	.012631	.0005148	24.53	0.000	.0116219	.01364
grade	.0783789	.0020912	37.48	0.000	.0742802	.0824777
not_smsa	-.1333091	.0089209	-14.94	0.000	-.1507938	-.1158243
south	-.1218994	.0121087	-10.07	0.000	-.145632	-.0981669
southXt	.0021033	.0008314	2.53	0.011	.0004738	.0037328
occ_code	-.0185603	.001033	-17.97	0.000	-.020585	-.0165355
_cons	.4567546	.032493	14.06	0.000	.3930695	.5204398
/sigma_u	.282881	.0038227	74.00	0.000	.2753886	.2903734
/sigma_e	.2696119	.0015957	168.96	0.000	.2664843	.2727394
rho	.524003	.0075625			.5091676	.5388052

Likelihood-ratio test of sigma_u=0: chibar2(01)= 6629.90 Prob>=chibar2 = 0.000

```
Observation summary:       14372      uncensored observations
                             157  left-censored observations
                             718 right-censored observations
                            3848      interval observations
```

The output includes the overall and panel-level variance components (labeled `sigma_e` and `sigma_u`, respectively) together with ρ (labeled `rho`),

$$\rho = \frac{\sigma_\nu^2}{\sigma_\epsilon^2 + \sigma_\nu^2}$$

which is the proportion of the total variance contributed by the panel-level variance component.

When `rho` is zero, the panel-level variance component is unimportant, and the panel estimator is not different from the pooled estimator. A likelihood-ratio test of this is included at the bottom of the output. This test formally compares the pooled estimator (intreg) with the panel estimator.

◁

❑ Technical Note

The random-effects model is calculated using quadrature. As the panel sizes (or ρ) increase, the quadrature approximation becomes less accurate. We can use the `quadchk` command to see if changing the number of quadrature points affects the results. If the results do change, then the quadrature approximation is not accurate, and the results of the model should not be interpreted. See [XT] **quadchk** for details and [XT] **xtprobit** for an example.

❑

Saved Results

xtintreg saves in e():

Scalars

e(N)	number of observations	e(g_min)	smallest group size
e(N_g)	number of groups	e(g_avg)	average group size
e(N_unc)	number of uncensored observations	e(chi2)	χ^2
e(N_lc)	number of left-censored observations	e(chi2_c)	χ^2 for comparison test
e(N_rc)	number of right-censored observations	e(rho)	ρ
e(N_int)	number of interval observations	e(sigma_u)	panel-level standard deviation
e(df_m)	model degrees of freedom	e(sigma_e)	standard deviation of ϵ_{it}
e(ll)	log likelihood	e(N_cd)	number of completely determined obs.
e(ll_0)	log likelihood, constant-only model	e(n_quad)	number of quadrature points
e(g_max)	largest group size		

Macros

e(cmd)	xtintreg	e(chi2type)	Wald or LR; type of model χ^2 test
e(depvar)	names of dependent variables	e(chi2_ct)	Wald or LR; type of model χ^2 test
e(title)	title in estimation output		corresponding to e(chi2_c)
e(ivar)	variable denoting groups	e(distrib)	Gaussian; the distribution of the
e(wtype)	weight type		random effect
e(wexp)	weight expression	e(crittype)	optimization criterion
e(offset1)	offset	e(predict)	program used to implement predict

Matrices

e(b)	coefficient vector	e(V)	variance–covariance matrix of the
			estimators

Functions

e(sample)	marks estimation sample

Methods and Formulas

xtintreg is implemented as an ado-file.

Assuming a normal distribution, $N(0, \sigma_\nu^2)$, for the random effects ν_i, we have the joint (unconditional of ν_i) density of the observed data for the ith panel

(Continued on next page)

$$f\left\{(y_{1i1}, y_{2i1}), \ldots, (y_{1in_i}, y_{2in_i}) | \mathbf{x}_{1i}, \ldots, \mathbf{x}_{in_i}\right\} =$$

$$\int_{-\infty}^{\infty} \frac{e^{-\nu_i^2/2\sigma_\nu^2}}{\sqrt{2\pi}\sigma_\nu} \left\{\prod_{t=1}^{n_i} F(y_{1it}, y_{2it}, \mathbf{x}_{it}\boldsymbol{\beta} + \nu_i)\right\} d\nu_i$$

where

$$F(y_{1it}, y_{2it}, \Delta_{it}) = \begin{cases} \left(\sqrt{2\pi}\sigma_\epsilon\right)^{-1} e^{-(y_{1it}-\Delta_{it})^2/(2\sigma_\epsilon^2)} & \text{if } (y_{1it}, y_{2it}) \in C \\[2mm] \Phi\left(\frac{y_{2it}-\Delta_{it}}{\sigma_\epsilon}\right) & \text{if } (y_{1it}, y_{2it}) \in L \\[2mm] 1 - \Phi\left(\frac{y_{1it}-\Delta_{it}}{\sigma_\epsilon}\right) & \text{if } (y_{1it}, y_{2it}) \in R \\[2mm] \Phi\left(\frac{y_{2it}-\Delta_{it}}{\sigma_\epsilon}\right) - \Phi\left(\frac{y_{1it}-\Delta_{it}}{\sigma_\epsilon}\right) & \text{if } (y_{1it}, y_{2it}) \in I \end{cases}$$

where C is the set of noncensored observations ($y_{1it} = y_{2it}$ and both nonmissing), L is the set of left-censored observations (y_{1it} missing and y_{2it} nonmissing), R is the set of right-censored observations (y_{1it} nonmissing and y_{2it} missing), I is the set of interval observations ($y_{1it} < y_{2it}$ and both nonmissing), and $\Phi()$ is the cumulative normal distribution. We can approximate the integral with M-point Gauss–Hermite quadrature

$$\int_{-\infty}^{\infty} e^{-x^2} g(x) dx \approx \sum_{m=1}^{M} w_m^* g(a_m^*)$$

where the w_m^* denote the quadrature weights and the a_m^* denote the quadrature abscissas. The log-likelihood L is then calculated using the quadrature

$$L = \sum_{i=1}^{n} w_i \log f\left\{(y_{1i1}, y_{2i1}), \ldots, (y_{1in_i}, y_{2in_i}) | \mathbf{x}_{1i}, \ldots, \mathbf{x}_{in_i}\right\}$$

$$\approx \sum_{i=1}^{n} w_i \log\left\{\frac{1}{\sqrt{\pi}} \sum_{m=1}^{M} w_m^* \prod_{t=1}^{n_i} F\left(y_{1it}, y_{2it}, \mathbf{x}_{it}\boldsymbol{\beta} + \sqrt{2}\sigma_\nu a_m^*\right)\right\}$$

where w_i is the user-specified weight for panel i; if no weights are specified, $w_i = 1$.

The quadrature formula requires that the integrated function be well-approximated by a polynomial. As the number of time periods becomes large (as panel size gets large),

$$\prod_{t=1}^{n_i} F(y_{1it}, y_{2it}, \mathbf{x}_{it}\boldsymbol{\beta} + \nu_i)$$

is no longer well-approximated by a polynomial. As a general rule of thumb, you should use this quadrature approach only for small to moderate panel sizes (based on simulations, 50 is a reasonably safe upper bound). However, if the data really come from random-effects intreg and rho is not too large (less than, say, .3), then the panel size could be 500 and the quadrature approximation would still be fine. If the data are not random-effects intreg or rho is large (bigger than, say, .7), then the quadrature approximation may be poor for panel sizes larger than 10. The quadchk command should be used to investigate the applicability of the numeric technique used in this command.

References

Neuhaus, J. M. 1992. Statistical methods for longitudinal and clustered designs with binary responses. *Statistical Methods in Medical Research* 1: 249–273.

Pendergast, J. F., S. J. Gange, M. A. Newton, M. J. Lindstrom, M. Palta, and M. R. Fisher. 1996. A survey of methods for analyzing clustered binary response data. *International Statistical Review* 64: 89–118.

Also See

Complementary:	[XT] **quadchk**, [XT] **xtdata**, [XT] **xtdes**, [XT] **xtsum**, [XT] **xttab**, [R] **adjust**, [R] **lincom**, [R] **mfx**, [R] **nlcom**, [R] **predict**, [R] **predictnl**, [R] **test**, [R] **testnl**, [R] **vce**
Related:	[XT] **xtgee**, [XT] **xtreg**, [XT] **xtregar**, [XT] **xttobit**, [R] **tobit**
Background:	[U] **16.5 Accessing coefficients and standard errors**, [U] **23 Estimation and post-estimation commands**, [XT] **xt**

Title

> **xtivreg** — Instrumental variables and two-stage least squares for panel-data models

Syntax

GLS Random-effects model

> xtivreg *depvar* $\left[\ varlist_1\ \right]$ (*varlist₂* = *varlist*ᵢᵥ) $\left[\text{if } exp\right]$ $\left[\text{in } range\right]$
>
> $\left[\ ,\ \text{re } \underline{\text{ec2}}\text{sls nosa } \underline{\text{reg}}\text{ress i}(\textit{varname}) \ \underline{\text{th}}\text{eta } \underline{\text{sm}}\text{all first } \underline{\text{l}}\text{evel}(\#)\ \right]$

Between-effects model

> xtivreg *depvar* $\left[\ varlist_1\ \right]$ (*varlist₂* = *varlist*ᵢᵥ) $\left[\text{if } exp\right]$ $\left[\text{in } range\right]$
>
> , be $\left[\ \underline{\text{reg}}\text{ress i}(\textit{varname}) \ \underline{\text{sm}}\text{all first } \underline{\text{l}}\text{evel}(\#)\ \right]$

Fixed-effects model

> xtivreg *depvar* $\left[\ varlist_1\ \right]$ (*varlist₂* = *varlist*ᵢᵥ) $\left[\text{if } exp\right]$ $\left[\text{in } range\right]$
>
> , fe $\left[\ \underline{\text{reg}}\text{ress i}(\textit{varname}) \ \underline{\text{sm}}\text{all first } \underline{\text{l}}\text{evel}(\#)\ \right]$

First-differenced estimator

> xtivreg *depvar* $\left[\ varlist_1\ \right]$ (*varlist₂* = *varlist*ᵢᵥ) $\left[\text{if } exp\right]$ $\left[\text{in } range\right]$
>
> , fd $\left[\ \underline{\text{reg}}\text{ress } \underline{\text{sm}}\text{all } \underline{\text{no}}\text{constant first } \underline{\text{l}}\text{evel}(\#)\ \right]$

by ... : may be used with xtivreg; see [R] **by**.

You must tsset your data before using xtivreg, fd; see [TS] **tsset**.

depvar, *varlist₁*, *varlist₂*, and *varlist*ᵢᵥ may contain time-series operators; see [U] **14.4.3 Time-series varlists**.

xtivreg shares the features of all estimation commands; see [U] **23 Estimation and post-estimation commands**.

(Continued on next page)

115

Syntax for predict

For all but the first-differenced estimator

> predict [*type*] *newvarname* [if *exp*] [in *range*] [, *statistic*]

where *statistic* is

xb	$\mathbf{Z}_{it}\widehat{\boldsymbol{\delta}}$, fitted values (the default)
ue	$\widehat{\mu}_i + \widehat{\nu}_{it}$, the combined residual
* xbu	$\mathbf{Z}_{it}\widehat{\boldsymbol{\delta}} + \widehat{\mu}_i$, prediction including effect
* u	$\widehat{\mu}_i$, the fixed or random error component
* e	$\widehat{\nu}_{it}$, the overall error component

Unstarred statistics are available both in and out of sample; type predict ... if e(sample) ... if wanted only for the estimation sample. Starred statistics are calculated only for the estimation sample even when if e(sample) is not specified.

First-differenced estimator

> predict [*type*] *newvarname* [if *exp*] [in *range*] [, *statistic*]

where *statistic* is

xb	$\mathbf{x}_j\mathbf{b}$, fitted values for the first-differenced model (the default)
e	$e_{it} - e_{it-1}$, the first-differenced overall error component

These statistics are available both in and out of sample; type predict ... if e(sample) ... if wanted only for the estimation sample.

Description

xtivreg offers five different estimators for panel-data models in which some of the right-hand-side covariates are endogenous. All five of the estimators are two-stage least-squares generalizations of simple panel-data estimators for the case of exogenous variables. xtivreg with the be option requests the two-stage least-squares between estimator. xtivreg with the fe option requests the two-stage least-squares within estimator. xtivreg with the re option requests a two-stage least-squares random-effects estimator. There are two implementations, G2SLS due to Balestra and Varadharajan-Krishnakumar (1987) and EC2SLS due to Baltagi. Since the Balestra and Varadharajan-Krishnakumar G2SLS is computationally less expensive, it is the default. Baltagi's EC2SLS can be obtained by specifying the additional ec2sls option. xtivreg with the fd option requests the two-stage least-squares first-differenced estimator.

See Baltagi (2001) for an introduction to panel-data models with endogenous covariates. For the derivation and application of the first-differenced estimator, see Anderson and Hsiao (1981).

Options

re requests the GLS random-effects estimator. re is the default.

be requests the between regression estimator.

fe requests the fixed-effects (within) regression estimator.

fd requests the first-differenced regression estimator.

ec2sls requests Baltagi's EC2SLS random-effects estimator instead of the default Balestra and Varadharajan-Krishnakumar estimator.

nosa specifies that the Baltagi–Chang estimators of the variance components are to be used instead of the default adapted Swamy–Arora estimators.

regress specifies that all the covariates are to be treated as exogenous and that the instrument list is to be ignored. In other words, specifying regress causes xtivreg to fit the requested panel-data regression model of *depvar* on *varlist*$_1$ and *varlist*$_2$, ignoring *varlist*$_{iv}$.

i(*varname*) specifies the variable name that contains the unit to which the observation belongs. You can specify the i() option the first time you estimate, or you can use the iis command to set i() beforehand. Note that it is not necessary to specify i() if the data have been previously tsset, or if iis has been previously specified—in these cases, the group variable is taken from the previous setting. See [XT] **xt**.

theta, used with xtreg, re only, specifies that the output should include the estimated value of θ used in combining the between and fixed estimators. For balanced data, this is a constant, and for unbalanced data, a summary of the values is presented in the header of the output.

small specifies that t statistics should be reported instead of Z statistics, and that F statistics should be reported instead of chi-squared statistics.

noconstant suppresses the constant term (intercept) in the regression.

first asks for the first-stage regressions to be displayed.

level(#) specifies the confidence level, in percent, for confidence intervals. The default is level(95) or as set by set level; see [U] **23.6 Specifying the width of confidence intervals**.

Options for predict

xb, the default, calculates the linear prediction; that is, $\mathbf{Z}_{it}\widehat{\boldsymbol{\delta}}$.

ue calculates the prediction of $\widehat{\mu}_i + \widehat{\nu}_{it}$. This is not available after the first-differenced model.

xbu calculates the prediction of $\mathbf{Z}_{it}\widehat{\boldsymbol{\delta}} + \widehat{\mu}_i$, the prediction including the fixed or random component. This is not available after the first-differenced model.

u calculates the prediction of $\widehat{\mu}_i$, the estimated fixed or random effect. This is not available after the first-differenced model.

e calculates the prediction of $\widehat{\nu}_{it}$.

Remarks

If you have not read [XT] **xt**, please do so.

Consider an equation of the form

$$y_{it} = \mathbf{Y}_{it}\boldsymbol{\gamma} + \mathbf{X}_{1it}\boldsymbol{\beta} + \mu_i + \nu_{it} = \mathbf{Z}_{it}\boldsymbol{\delta} + \mu_i + \nu_{it} \qquad (1)$$

where

y_{it} is the dependent variable

$\mathbf{Y}_{it}$ is an $1 \times g_2$ vector of observations on g_2 endogenous variables included as covariates, and these variables are allowed to be correlated with the ν_{it}

$\mathbf{X}_{1it}$ is an $1 \times k_1$ vector of observations on the exogenous variables included as covariates

$\mathbf{Z}_{it} = [\mathbf{Y}_{it} \ \mathbf{X}_{it}]$

γ is a $g_2 \times 1$ vector of coefficients

β is a $k_1 \times 1$ vector of coefficients

δ is a $K \times 1$ vector of coefficients, and $K = g_2 + k_1$

Assume that there is a $1 \times k_2$ vector of observations on the k_2 instruments in $\mathbf{X}_{2it}$. The order condition is satisfied if $k_2 \geq g_2$. Let $\mathbf{X}_{it} = [\mathbf{X}_{1it} \ \mathbf{X}_{2it}]$. xtivreg handles exogenously unbalanced panel data. Thus, define T_i to be the number of observations on panel i, n to be the number of panels, and N to be the total number of observations; i.e., $N = \sum_{i=1}^{n} T_i$.

xtivreg offers five different estimators, which may be applied to models of the equation (1) form. The first-differenced estimator (FD2SLS) removes the μ_i by fitting the model in first differences. The within estimator (FE2SLS) fits the model after sweeping out the μ_i by removing the panel-level means from each variable. The between estimator (BE2SLS) models the panel averages. The two random-effects estimators, G2SLS and EC2SLS, treat the μ_i as random variables that are independent and identically distributed (i.i.d.) over the panels. Except for (FD2SLS), all of these estimators are generalizations of estimators in xtreg. See [XT] **xtreg** for a discussion of these estimators in the case of exogenous covariates.

While the estimators allow for different assumptions about the μ_i, all of the estimators assume that the idiosyncratic error term ν_{it} has zero mean and is uncorrelated with the variables in $\mathbf{X}_{it}$. Just as in the case where there are no endogenous covariates, as discussed in [XT] **xtreg**, there are varying perspectives on what assumptions should be placed on the μ_i. The μ_i may be assumed to be fixed or random. If they are assumed to be fixed, then the μ_i may be correlated with the variables in $\mathbf{X}_{it}$, and the within estimator is efficient within a class of limited information estimators. Alternatively, the μ_i may be assumed to be random. In this case, the μ_i are assumed to be independent and identically distributed over the panels. If the μ_i are assumed to be uncorrelated with the variables in $\mathbf{X}_{it}$, then the GLS random-effects estimators are more efficient than the within estimator. However, if the μ_{it} are correlated with the variables in $\mathbf{X}_{it}$, then the random-effects estimators are inconsistent but the within estimator is consistent. The price of using the within estimator is that it is not possible to estimate coefficients on time-invariant variables, and all inference is conditional on the μ_i in the sample. See Mundlak (1978) and Hsiao (1986) for discussions of this interpretation of the within estimator.

▷ Example

The two-stage least-squares first-differenced estimator (FD2SLS) has been used to fit both "fixed-effect" and "random-effect" models. If the μ_i are truly fixed-effects, then the FD2SLS estimator is not as efficient as the two-stage least-squares within estimator for finite T_i. Similarly, if none of the endogenous variables are lagged dependent variables, the exogenous variables are all strictly exogenous, and the random effects are i.i.d. and independent of the $\mathbf{X}_{it}$, then the two-stage GLS estimators are more efficient than the FD2SLS estimator. However, the FD2SLS estimator has been used to obtain consistent estimates when one of these conditions fails. Perhaps most notably, Anderson and Hsiao (1981) used a version of the FD2SLS estimator to fit a panel-data model with a lagged dependent variable.

Arellano and Bond (1991) developed new one-step and two-step GMM estimators for dynamic panel data. See [XT] **xtabond** for a discussion of these estimators and Stata's implementation of them. In their article, Arellano and Bond (1991) applied their new estimators to a model of dynamic labor demand that had previously been considered by Layard and Nickell (1986). They also compared the results of their estimators with those from the Anderson–Hsiao estimator. They used data from an unbalanced panel of firms from the United Kingdom. As is conventional, all variables are indexed over the firm i and time t. In this dataset, n_{it} is the log of employment in firm i inside the U.K. at time t, w_{it} is the natural log of the real product wage, k_{it} is the natural log of the gross capital stock, and ys_{it} is the natural log of industry output. The model also includes time dummies yr1980, yr1981, yr1982, yr1983, and yr1984. In Arellano and Bond (1991, Table 5, column e), the authors present the results they obtained from applying one version of the Anderson–Hsiao estimator to these data. This example reproduces their results for the coefficients. The standard errors are different because Arellano and Bond are using robust standard errors.

```
. use http://www.stata-press.com/data/r8/abdata

. xtivreg n l2.n l(0/1).w l(0/2).(k ys) yr1981-yr1984 (l.n = l3.n), fd

First-differenced IV regression            Number of obs      =        471
Group variable: id                         Number of groups   =        140

R-sq:  within  = 0.0141                     Obs per group: min =          3
       between = 0.9165                                    avg =        3.4
       overall = 0.9892                                    max =          5

                                           chi2(14)           =     122.53
corr(u_i, Xb)  = 0.9239                     Prob > chi2        =     0.0000
```

d.n		Coef.	Std. Err.	z	P>|z|	[95% Conf. Interval]	
n							
	LD	1.422765	1.583053	0.90	0.369	-1.679962	4.525493
	L2D	-.1645517	.1647179	-1.00	0.318	-.4873928	.1582894
w							
	D1	-.7524675	.1765733	-4.26	0.000	-1.098545	-.4063902
	LD	.9627611	1.086506	0.89	0.376	-1.166752	3.092275
k							
	D1	.3221686	.1466086	2.20	0.028	.0348211	.6095161
	LD	-.3248778	.5800599	-0.56	0.575	-1.461774	.8120187
	L2D	-.0953947	.1960883	-0.49	0.627	-.4797207	.2889314
ys							
	D1	.7660906	.369694	2.07	0.038	.0415037	1.490678
	LD	-1.361881	1.156835	-1.18	0.239	-3.629237	.9054744
	L2D	.3212993	.5440403	0.59	0.555	-.745	1.387599
yr1981							
	D1	-.0574197	.0430158	-1.33	0.182	-.1417291	.0268896
yr1982							
	D1	-.0882952	.0706214	-1.25	0.211	-.2267106	.0501203
yr1983							
	D1	-.1063153	.10861	-0.98	0.328	-.319187	.1065563
yr1984							
	D1	-.1172108	.15196	-0.77	0.441	-.4150468	.1806253
_cons		.0161204	.0336264	0.48	0.632	-.0497861	.082027

sigma_u	.29069213	
sigma_e	.18855982	
rho	.70384993	(fraction of variance due to u_i)

```
Instrumented:  L.n
Instruments:   L2.n w L.w k L.k L2.k ys L.ys L2.ys yr1981 yr1982 yr1983 yr1984
               L3.n
```

▷ Example

For the within estimator, let's consider another version of the wage equation discussed in [XT] **xtreg**. The data for this example come from an extract of women from the National Longitudinal Survey of Youth that was described in detail in [XT] **xt**. Restricting ourselves to only time-varying covariates, we might suppose that the log of the real wage was a function of the individuals' age, age^2, her tenure in the observed place of employment, whether or not she belonged to union, whether or not she lives in metropolitan area, and whether or not she lives in the south. The variables for these are, respectively, age, age2, tenure, union, not_smsa, and south. If we treat all the variables as exogenous, then we could use the one-stage within estimator from xtreg. This would yield

```
. use http://www.stata-press.com/data/r8/nlswork
(National Longitudinal Survey.  Young Women 14-26 years of age in 1968)

. generate age2 = age^2
(24 missing values generated)

. xtreg ln_w age* tenure not_smsa union south, fe i(idcode)
```

Fixed-effects (within) regression	Number of obs	=	19007
Group variable (i): idcode	Number of groups	=	4134

R-sq: within = 0.1333	Obs per group: min =	1
between = 0.2375	avg =	4.6
overall = 0.2031	max =	12

	F(6,14867)	=	381.19
corr(u_i, Xb) = 0.2074	Prob > F	=	0.0000

ln_wage	Coef.	Std. Err.	t	P>|t|	[95% Conf. Interval]	
age	.0311984	.0033902	9.20	0.000	.0245533	.0378436
age2	-.0003457	.0000543	-6.37	0.000	-.0004522	-.0002393
tenure	.0176205	.0008099	21.76	0.000	.0160331	.0192079
not_smsa	-.0972535	.0125377	-7.76	0.000	-.1218289	-.072678
union	.0975672	.0069844	13.97	0.000	.0838769	.1112576
south	-.0620932	.013327	-4.66	0.000	-.0882158	-.0359706
_cons	1.091612	.0523126	20.87	0.000	.9890729	1.194151

sigma_u	.3910683	
sigma_e	.25545969	
rho	.70091004	(fraction of variance due to u_i)

F test that all u_i=0:	F(4133,14867) =	8.31	Prob > F = 0.0000

All the coefficients are statistically significant and have the expected signs.

Now, suppose that we wish to model tenure as a function of union and south, and that we believe that the errors in the two equations are correlated. Since we are still interested in the within estimates, we now need a two-stage least-squares estimator. The following output shows the command and the results from fitting this model:

(Continued on next page)

```
. xtivreg ln_w age* not_smsa (tenure = union south), fe i(idcode)
Fixed-effects (within) IV regression         Number of obs       =       19007
Group variable: idcode                       Number of groups    =        4134

R-sq:  within  =      .                       Obs per group: min =           1
       between = 0.1304                                       avg =         4.6
       overall = 0.0897                                       max =          12

                                             Wald chi2(4)        =   147926.58
corr(u_i, Xb)  = -0.6843                      Prob > chi2         =      0.0000
```

ln_wage	Coef.	Std. Err.	z	P>|z|	[95% Conf. Interval]	
tenure	.2403531	.0373419	6.44	0.000	.1671643	.3135419
age	.0118437	.0090032	1.32	0.188	-.0058023	.0294897
age2	-.0012145	.0001968	-6.17	0.000	-.0016003	-.0008286
not_smsa	-.0167178	.0339236	-0.49	0.622	-.0832069	.0497713
_cons	1.678287	.1626657	10.32	0.000	1.359468	1.997106

sigma_u	.70661941	
sigma_e	.63029359	
rho	.55690561	(fraction of variance due to u_i)

```
F test that all u_i=0:      F(4133,14869) =      1.44      Prob > F    = 0.0000
```

```
Instrumented:   tenure
Instruments:    age age2 not_smsa union south
```

Although all the coefficients still have the expected signs, the coefficients on age and not_smsa are no longer statistically significant. Given that these variables have been found to be important in many other studies, we might want to rethink our specification.

◁

Now, suppose that we are willing to assume that the μ_i are uncorrelated with the other covariates, so we can fit a random-effects model. The model is frequently known as the variance components or error components model. xtivreg has estimators for two-stage least-squares one-way error component models. In the one-way framework, there are two variance components to estimate, the variance of the μ_i and the component for the ν_{it}. Since the variance components are unknown, consistent estimates are required to implement feasible GLS. xtivreg offers two choices: a Swamy–Arora method and simple consistent estimators due to Baltagi and Chang (2000).

Baltagi and Chang (1994) derived the Swamy–Arora estimators of the variance components for unbalanced panels. By default, xtivreg uses estimators that extend these unbalanced Swamy–Arora estimators to the case with instrumental variables. The default Swamy–Arora method contains a degree-of-freedom correction to improve its performance in small samples. Baltagi and Chang (2000) used variance-components estimators, which are based on the ideas of Amemiya (1971) and Swamy and Arora (1972), but do not attempt to make small-sample adjustments. These consistent estimators of the variance components will be used if the nosa option is specified.

Using either estimator of the variance components, xtivreg offers two GLS estimators of the random-effects model. These two estimators differ only in how they construct the GLS instruments from the exogenous and instrumental variables contained in $\mathbf{X}_{it} = [\mathbf{X}_{1it}\ \mathbf{X}_{2it}]$. The default method, G2SLS, which is due to Balestra and Varadharajan-Krishnakumar, uses the exogenous variables after they have been passed through the feasible GLS transform. In math, G2SLS uses $\mathbf{X}_{it}^*$ for the GLS instruments, where $\mathbf{X}_{it}^*$ is constructed by passing each variable in $\mathbf{X}_{it}$ through the GLS transform in (3), given in the *Methods and Formulas* section. If the ec2sls option is specified, then xtivreg performs Baltagi's EC2SLS. In EC2SLS, the instruments are $\widetilde{\mathbf{X}}_{it}$ and $\overline{\mathbf{X}}_{it}$, where $\widetilde{\mathbf{X}}_{it}$ is constructed by passing each of the variables in $\mathbf{X}_{it}$ through the within transform, and $\overline{\mathbf{X}}_{it}$ is constructed by passing each variable through the between transform. The within and between transforms are given in the *Methods and Formulas* section. Baltagi and Li (1992) showed that although the G2SLS instruments are a subset of those contained in EC2SLS, the extra instruments in EC2SLS are redundant in the sense of White (1984). Given the extra computational cost, G2SLS is the default.

▷ Example

Here is the output from applying the G2SLS estimator to this model:

```
. generate byte black = (race==2)

. xtivreg ln_w age* not_smsa black (tenure = union birth south black), re
> i(idcode)
```

| G2SLS random-effects IV regression | | | | Number of obs | = | 19007 |

G2SLS random-effects IV regression   Number of obs   =   19007
Group variable: idcode   Number of groups =   4134

R-sq: within = 0.0664   Obs per group: min =   1
       between = 0.2098   avg =   4.6
       overall = 0.1463   max =   12

                                              Wald chi2(5)   =   1446.37
corr(u_i, X)   = 0 (assumed)   Prob > chi2   =   0.0000

ln_wage	Coef.	Std. Err.	z	P>|z|	[95% Conf. Interval]	
tenure	.1391798	.0078756	17.67	0.000	.123744	.1546157
age	.0279649	.0054182	5.16	0.000	.0173454	.0385843
age2	-.0008357	.0000871	-9.60	0.000	-.0010063	-.000665
not_smsa	-.2235103	.0111371	-20.07	0.000	-.2453386	-.2016821
black	-.2078613	.0125803	-16.52	0.000	-.2325183	-.1832044
_cons	1.337684	.0844988	15.83	0.000	1.172069	1.503299
sigma_u	.36582493					
sigma_e	.63031479					
rho	.25197078	(fraction of variance due to u_i)				

```
Instrumented:  tenure
Instruments:   age age2 not_smsa black union birth_yr south
```

Note that we have included two time-invariant covariates, birth_yr and black. All the coefficients are statistically significant and are of the expected sign.

Applying the EC2SLS estimator yields similar results:

(Continued on next page)

```
. xtivreg ln_w age* not_smsa black (tenure = union birth south black), re
> ec2sls i(idcode)
```

```
EC2SLS random-effects IV regression          Number of obs      =     19007
Group variable: idcode                        Number of groups   =      4134

R-sq:  within  = 0.0898                        Obs per group: min =         1
       between = 0.2608                                       avg =       4.6
       overall = 0.1926                                       max =        12

                                              Wald chi2(5)       =   2721.92
corr(u_i, X)       = 0 (assumed)              Prob > chi2        =    0.0000
```

ln_wage	Coef.	Std. Err.	z	P>|z|	[95% Conf. Interval]	
tenure	.064822	.0025647	25.27	0.000	.0597953	.0698486
age	.0380048	.0039549	9.61	0.000	.0302534	.0457562
age2	-.0006676	.0000632	-10.56	0.000	-.0007915	-.0005438
not_smsa	-.2298961	.0082993	-27.70	0.000	-.2461625	-.2136297
black	-.1823627	.0092005	-19.82	0.000	-.2003954	-.16433
_cons	1.110564	.0606538	18.31	0.000	.9916849	1.229443

sigma_u	.36582493			
sigma_e	.63031479			
rho	.25197078	(fraction of variance due to u_i)		

```
Instrumented:   tenure
Instruments:    age age2 not_smsa black union birth_yr south
```

Fitting the same model as above with the G2SLS estimator and the consistent variance components estimators yields

```
. xtivreg ln_w age* not_smsa black (tenure = union birth south black), re
> nosa i(idcode)
```

```
G2SLS random-effects IV regression            Number of obs      =     19007
Group variable: idcode                        Number of groups   =      4134

R-sq:  within  = 0.0664                        Obs per group: min =         1
       between = 0.2098                                       avg =       4.6
       overall = 0.1463                                       max =        12

                                              Wald chi2(5)       =   1446.93
corr(u_i, X)       = 0 (assumed)              Prob > chi2        =    0.0000
```

ln_wage	Coef	Std. Err.	z	P>|z|	[95% Conf. Interval]	
tenure	.1391859	.007873	17.68	0.000	37552	.1546166
age	.0279697	.005419	5.16	0.000	.0173486	.0385909
age2	-.0008357	.0000871	-9.60	0.000	-.0010064	-.000665
not_smsa	-.2235738	.0111344	-20.08	0.000	-.2453967	-.2017508
black	-.2078733	.0125751	-16.53	0.000	-.2325201	-.1832265
_cons	1.337522	.0845083	15.83	0.000	1.171889	1.503155

sigma_u	.36535633			
sigma_e	.63020883			
rho	.2515512	(fraction of variance due to u_i)		

```
Instrumented:   tenure
Instruments:    age age2 not_smsa black union birth_yr south
```

◁

Acknowledgment

We thank Mead Over of the World Bank, who wrote an early implementation of xtivreg.

Saved Results

xtivreg, re saves in e():

Scalars

e(N)	number of observations	e(r2_o)	R-squared for overall model
e(N_g)	number of groups	e(r2_b)	R-squared for between model
e(df_m)	model degrees of freedom	e(sigma)	ancillary parameter (gamma, lnormal)
e(g_max)	largest group size	e(sigma_u)	panel-level standard deviation
e(g_min)	smallest group size	e(sigma_e)	standard deviation of ϵ_{it}
e(g_avg)	average group size	e(thta_min)	minimum θ
e(chi2)	χ^2	e(thta_5)	θ, 5th percentile
e(rho)	ρ	e(thta_50)	θ, 50th percentile
e(Tbar)	harmonic mean of group sizes	e(thta_95)	θ, 95th percentile
e(F)	model F (small only)	e(thta_max)	maximum θ
e(df_rz)	residual degrees of freedom	e(m_p)	p-value from model test
e(r2_w)	R-squared for within model		

Macros

e(cmd)	xtivreg	e(insts)	instruments
e(depvar)	name of dependent variable	e(instd)	instrumented variables
e(model)	g2sls or ec2sls	e(chi2type)	Wald; type of model χ^2 test
e(ivar)	variable denoting groups	e(predict)	program used to implement predict

Matrices

e(b)	coefficient vector	e(V)	variance–covariance matrix of the estimators

Functions

e(sample)	marks estimation sample

(*Continued on next page*)

`xtivreg, be` saves in `e()`:

Scalars

e(N)	number of observations	e(g_max)	largest group size
e(N_g)	number of groups	e(g_min)	smallest group size
e(df_m)	model degrees of freedom	e(g_avg)	average group size
e(rss)	residual sum of squares	e(Tcon)	1 if T is constant
e(df_r)	residual degrees of freedom	e(r2)	R-squared
e(chi2)	model Wald	e(r2_w)	R-squared for within model
e(F)	F statistic (small only)	e(r2_o)	R-squared for overall model
e(rmse)	root mean square error	e(r2_b)	R-squared for between model

Macros

e(cmd)	xtivreg	e(insts)	instruments
e(depvar)	name of dependent variable	e(instd)	instrumented variables
e(model)	be	e(small)	small if specified
e(ivar)	variable denoting groups	e(predict)	program used to implement predict

Matrices

e(b)	coefficient vector	e(V)	variance–covariance matrix of the estimators

Functions

e(sample)	marks estimation sample

`xtivreg, fe` saves in `e()`:

Scalars

e(N)	number of observations	e(g_max)	largest group size
e(N_g)	number of groups	e(g_min)	smallest group size
e(mss)	model sum of squares	e(g_avg)	average group size
e(tss)	total sum of squares	e(rho)	ρ
e(df_m)	model degrees of freedom	e(Tbar)	harmonic mean of group sizes
e(rss)	residual sum of squares	e(Tcon)	1 if T is constant
e(df_r)	residual d.o.f. (small only)	e(r2_w)	R-squared for within model
e(r2)	R-squared	e(r2_o)	R-squared for overall model
e(r2_a)	adjusted R-squared	e(r2_b)	R-squared for between model
e(F)	F statistic (small only)	e(sigma)	ancillary parameter (gamma, lnormal)
e(rmse)	root mean square error	e(corr)	corr(u_i, Xb)
e(chi2)	model Wald (not small)	e(sigma_u)	panel-level standard deviation
e(df_a)	degrees of freedom for absorbed effect	e(sigma_e)	standard deviation of ϵ_{it}
e(F_f)	F for H_0: $u_i=0$		

Macros

e(cmd)	xtivreg	e(insts)	instruments
e(depvar)	name of dependent variable	e(instd)	instrumented variables
e(model)	fe	e(predict)	program used to implement predict
e(ivar)	variable denoting groups		

Matrices

e(b)	coefficient vector	e(V)	variance–covariance matrix of the estimators

Functions

e(sample)	marks estimation sample

`xtivreg, fd` saves in `e()`:

Scalars

`e(N)`	number of observations	`e(g_max)`	largest group size
`e(N_g)`	number of groups	`e(g_min)`	smallest group size
`e(mss)`	model sum of squares	`e(g_avg)`	average group size
`e(tss)`	total sum of squares	`e(rho)`	ρ
`e(df_m)`	model degrees of freedom	`e(Tbar)`	harmonic mean of group sizes
`e(rss)`	residual sum of squares	`e(Tcon)`	1 if T is constant
`e(df_r)`	residual d.o.f. (small only)	`e(r2_w)`	R-squared for within model
`e(r2)`	R-squared	`e(r2_o)`	R-squared for overall model
`e(r2_a)`	adjusted R-squared	`e(r2_b)`	R-squared for between model
`e(F)`	F statistic (small only)	`e(sigma)`	ancillary parameter (gamma, lnormal)
`e(rmse)`	root mean square error	`e(corr)`	$\mathrm{corr}(u_i, \mathrm{Xb})$
`e(chi2)`	model Wald (not small)	`e(sigma_u)`	panel-level standard deviation
`e(df_a)`	degrees of freedom for absorbed effect	`e(sigma_e)`	standard deviation of ϵ_{it}
`e(F_f)`	F for H_0: $u_i = 0$		

Macros

`e(cmd)`	xtivreg	`e(tvar)`	time variable
`e(depvar)`	name of dependent variable	`e(insts)`	instruments
`e(model)`	fd	`e(instd)`	instrumented variables
`e(ivar)`	variable denoting groups	`e(predict)`	program used to implement predict

Matrices

`e(b)`	coefficient vector	`e(V)`	variance–covariance matrix of the estimators

Functions

`e(sample)`	marks estimation sample

Methods and Formulas

Consider an equation of the form

$$y_{it} = \mathbf{Y}_{it}\gamma + \mathbf{X}_{1it}\beta + \mu_i + \nu_{it} = \mathbf{Z}_{it}\delta + \mu_i + \nu_{it} \tag{2}$$

where

y_{it} is the dependent variable

$\mathbf{Y}_{it}$ is an $1 \times g_2$ vector of observations on g_2 endogenous variables included as covariates, and these variables are allowed to be correlated with the ν_{it}

$\mathbf{X}_{1it}$ is an $1 \times k_1$ vector of observations on the exogenous variables included as covariates

$\mathbf{Z}_{it} = [\mathbf{Y}_{it}\ \mathbf{X}_{it}]$

γ is a $g_2 \times 1$ vector of coefficients

β is a $k_1 \times 1$ vector of coefficients

δ is a $K \times 1$ vector of coefficients, where $K = g_2 + k_1$

Assume that there is a $1 \times k_2$ vector of observations on the k_2 instruments in $\mathbf{X}_{2it}$. The order condition is satisfied if $k_2 \geq g_2$. Let $\mathbf{X}_{it} = [\mathbf{X}_{1it}\ \mathbf{X}_{2it}]$. xtivreg handles exogenously unbalanced panel data. Thus, define T_i to be the number of observations on panel i, n to be the number of panels, and N to be the total number of observations; i.e., $N = \sum_{i=1}^{n} T_i$.

xtivreg, fd

As the name implies, this estimator obtains its estimates from an instrumental-variables regression on the first-differenced data. Specifically, first differencing the data yields

$$y_{it} - y_{it-1} = (\mathbf{Z}_{it} - \mathbf{Z}_{i,t-1})\,\delta + \nu_{it} - \nu_{i,t-1}$$

With the μ_i removed by differencing, we can obtain the estimated coefficients and their estimated variance–covariance matrix from a standard two-stage least-squares regression of Δy_{it} on $\Delta \mathbf{Z}_{it}$ with instruments $\Delta \mathbf{X}_{it}$.

Reported as R^2 within is $\left[\mathrm{corr}\big\{(\mathbf{Z}_{it} - \overline{\mathbf{Z}}_i)\widehat{\boldsymbol{\delta}}, y_{it} - \overline{y}_i\big\}\right]^2$.

Reported as R^2 between is $\left\{\mathrm{corr}(\overline{\mathbf{Z}}_i\widehat{\boldsymbol{\delta}}, \overline{y}_i)\right\}^2$.

Reported as R^2 overall is $\left\{\mathrm{corr}(\mathbf{Z}_{it}\widehat{\boldsymbol{\delta}}, y_{it})\right\}^2$.

xtreg, fe

At the heart of this model is the within transformation. The within transform of a variable w is

$$\widetilde{w}_{it} = w_{it} - \overline{w}_{i.} + \overline{w}$$

where

$$\overline{w}_{i.} = \left(\frac{1}{n}\right)\sum_{t=1}^{T_i} w_{it}$$

$$\overline{w} = \left(\frac{1}{N}\right)\sum_{i=1}^{n}\sum_{t=1}^{T_i} w_{it}$$

and n is the number of groups and N is the total number of observations on the variable.

The within transform of (2) is

$$\widetilde{y}_{it} = \widetilde{\mathbf{Z}}_{it} + \widetilde{\nu}_{it}$$

Note that the within transform has removed the μ_i. With the μ_i gone, the within 2SLS estimator can be obtained from a two-stage least-squares regression of $\widetilde{y}_{it}$ on $\widetilde{\mathbf{Z}}_{it}$ with instruments $\widetilde{\mathbf{X}}_{it}$.

Suppose that there are K variables in $\mathbf{Z}_{it}$, including the mandatory constant. Then, there are $K + n - 1$ parameters estimated in the model, and the VCE for the within estimator is

$$\frac{N - K}{N - n - K + 1} V_{IV}$$

where V_{IV} is the VCE from the above two-stage least-squares regression.

From the estimate of $\widehat{\boldsymbol{\delta}}$, estimates $\widehat{\mu}_i$ of μ_i are obtained as $\widehat{\mu}_i = \overline{y}_i - \overline{\mathbf{Z}}_i\widehat{\boldsymbol{\delta}}$. Reported from the calculated $\widehat{\mu}_i$ is its standard deviation and its correlation with $\overline{\mathbf{Z}}_i\widehat{\boldsymbol{\delta}}$. Reported as the standard deviation of ν_{it} is the regression's estimated root mean square error, s^2, which is adjusted (as previously stated) for the $n - 1$ estimated means.

Reported as R^2 within is the R^2 from the mean-deviated regression.

Reported as R^2 between is $\left\{ \mathrm{corr}(\overline{\mathbf{Z}}_i \widehat{\boldsymbol{\delta}}, \overline{y}_i) \right\}^2$.

Reported as R^2 overall is $\left\{ \mathrm{corr}(\mathbf{Z}_{it} \widehat{\boldsymbol{\delta}}, y_{it}) \right\}^2$.

At the bottom of the output, an F statistic against the null hypothesis that all the μ_i are zero is reported. This F statistic is an application of the results in Wooldridge (1990).

xtivreg, be

After passing (2) through the between transform, we are left with

$$\overline{y}_i = \alpha + \overline{\mathbf{Z}}_i \boldsymbol{\delta} + \mu_i + \overline{\nu}_i \tag{3}$$

where

$$\overline{w}_i = \left(\frac{1}{T_i} \right) \sum_{t=1}^{T_i} w_{it} \quad \text{for } w \in \{y, \mathbf{Z}, \nu\}$$

Similarly, define $\overline{\mathbf{X}}_i$ as the matrix of instruments $\mathbf{X}_{it}$ after they have been passed through the between transform.

The BE2SLS estimator of (3) obtains its coefficient estimates and its VCE, a two-stage least-squares regression of $\overline{y}_i$ on $\overline{\mathbf{Z}}_i$ with instruments $\overline{\mathbf{X}}_i$ in which each average appears T_i times.

Reported as R^2 between is the R^2 from the fitted regression.

Reported as R^2 within is $\left[\mathrm{corr}\left\{ (\mathbf{Z}_{it} - \overline{\mathbf{Z}}_i)\widehat{\boldsymbol{\delta}}, y_{it} - \overline{y}_i \right\} \right]^2$.

Reported as R^2 overall is $\left\{ \mathrm{corr}(\mathbf{Z}_{it} \widehat{\boldsymbol{\delta}}, y_{it}) \right\}^2$.

xtreg, re

Following Baltagi and Chang (2000), let

$$u = \mu_i + \nu_{it}$$

be the $N \times 1$ vector of combined errors. Then, under the assumptions of the random-effects model,

$$E(uu') = \sigma_\nu^2 \mathrm{diag}\left[I_{T_i} - \frac{1}{T_i} \iota_{T_i} \iota'_{T_i} \right] + \mathrm{diag}\left[w_i \frac{1}{T_i} \iota_{T_i} \iota'_{T_i} \right]$$

where

$$\omega_i = T_i \sigma_\mu^2 + \sigma_\nu^2$$

and ι_{T_i} is a vector of ones of dimension T_i.

Since the variance components are unknown, consistent estimates are required to implement feasible GLS. xtivreg offers two choices. The default is a simple extension of the Swamy–Arora method for unbalanced panels.

Let

$$u_{it}^w = \widetilde{y}_{it} - \widetilde{\mathbf{Z}}_{it} \widehat{\boldsymbol{\delta}}_w$$

be the combined residuals from the within estimator. Let $\widetilde{u}_{it}$ be the within transformed u_{it}. Then,

$$\widehat{\sigma}_\nu = \frac{\sum_{i=1}^n \sum_{t=1}^{T_i} \widetilde{u}_{it}^2}{N - n - K + 1}$$

Let

$$u_{it}^b = y_{it} - \mathbf{Z}_{it}\boldsymbol{\delta}_b$$

be the combined residual from the between estimator. Let $\overline{u}_{i.}^b$ be the between residuals after they have been passed through the between transform. Then,

$$\widehat{\sigma}_\mu^2 = \frac{\sum_{i=1}^n \sum_{t=1}^{T_i} \overline{u}_{it}^2 - (n-K)\widehat{\sigma}_\nu^2}{N-r}$$

where

$$r = \text{trace}\left\{\left(\overline{\mathbf{Z}}_i'\overline{\mathbf{Z}}_i\right)^{-1}\overline{\mathbf{Z}}_i'\mathbf{Z}_\mu\mathbf{Z}_\mu'\overline{\mathbf{Z}}_i\right\}$$

where

$$\mathbf{Z}_\mu = \text{diag}\left(\iota_{T_i}\iota_{T_i}'\right)$$

If the `nosa` option is specified, then the consistent estimators described in Baltagi and Chang (2000) are used. These are given by

$$\widehat{\sigma}_\nu = \frac{\sum_{i=1}^n \sum_{t=1}^{T_i} \widetilde{u}_{it}^2}{N-n}$$

and

$$\widehat{\sigma}_\mu^2 = \frac{\sum_{i=1}^n \sum_{t=1}^{T_i} \overline{u}_{it}^2 - n\widehat{\sigma}_\nu^2}{N}$$

Note that the default Swamy–Arora method contains a degree of freedom correction to improve its performance in small samples.

Given estimates of the variance components, $\widehat{\sigma}_\nu^2$ and $\widehat{\sigma}_\mu^2$, the feasible GLS transform of a variable w is

$$w^* = w_{it} - \widehat{\theta}_{it}\overline{w}_{i.} \tag{4}$$

where

$$\overline{w}_{i.} = \left(\frac{1}{T_i}\right)\sum_{t=1}^{T_i} w_{it}$$

$$\widehat{\theta}_{it} = 1 - \left(\frac{\widehat{\sigma}_\nu^2}{\widehat{\omega}_i}\right)^{-\frac{1}{2}}$$

and

$$\widehat{\omega}_i = T_i\widehat{\sigma}_\mu^2 + \widehat{\sigma}_\nu^2$$

Using either estimator of the variance components, `xtivreg` contains two GLS estimators of the random-effects model. These two estimators differ only in how they construct the GLS instruments from the exogenous and instrumental variables contained in $\mathbf{X}_{it} = [\mathbf{X}_{1it}\mathbf{X}_{2it}]$. The default method, G2SLS, which is due to Balestra and Varadharajan-Krishnakumar, uses the exogenous variables after they have been passed through the feasible GLS transform. In math, G2SLS uses $\mathbf{X}^*$ for the GLS instruments, where $\mathbf{X}^*$ is constructed by passing each variable in $\mathbf{X}$ though the GLS transform in (4). The G2SLS estimator obtains its coefficient estimates and VCE from an instrumental variable regression of y_{it}^* on $\mathbf{Z}_{it}^*$ with instruments $\mathbf{X}_{it}^*$.

If the `ec2sls` option is specified, then `xtivreg` performs Baltagi's EC2SLS. In EC2SLS, the instruments are $\widetilde{\mathbf{X}}_i t$ and $\overline{\mathbf{X}}_{it}$, where $\widetilde{X}_{it}$ is constructed by each of the variables in $\mathbf{X}_{it}$ throughout the GLS transform in (4), and $\overline{\mathbf{X}}_{it}$ is made of the group means of each variable in $\mathbf{X}_{it}$. The EC2SLS estimator obtains its coefficient estimates and its VCE from an instrumental variables regression of y_{it}^* on $\mathbf{Z}_{it}^*$ with instruments $\widetilde{\mathbf{X}}_{it}$ and $\overline{bfX}_{it}$.

Baltagi and Li (1992) showed that although the G2SLS instruments are a subset of those in contained in EC2SLS, the extra instruments in EC2SLS are redundant in the sense of White (1984). Given the extra computational cost, G2SLS is the default.

The standard deviation of $\mu_i + \nu_{it}$ is calculated as $\sqrt{\widehat{\sigma}_\mu^2 + \widehat{\sigma}_\nu^2}$.

Reported as R^2 between is $\left\{ \mathrm{corr}(\overline{\mathbf{Z}}_i\widehat{\boldsymbol{\delta}}, \overline{y}_i) \right\}^2$.

Reported as R^2 within is $\left[\mathrm{corr}\left\{ (\mathbf{Z}_{it} - \overline{\mathbf{Z}}_i)\widehat{\boldsymbol{\delta}}, y_{it} - \overline{y}_i \right\} \right]^2$.

Reported as R^2 overall is $\left\{ \mathrm{corr}(\mathbf{Z}_{it}\widehat{\boldsymbol{\delta}}, y_{it}) \right\}^2$.

References

Amemiya, T. 1971. The estimation of the variances is a variance-components model. *International Economic Review* 12: 1–13.

Anderson, T. W. and C. Hsiao. 1981. Estimation of dynamic models with error components. *Journal of the American Statistical Association* 76: 598–606.

Balestra, P. and J. Varadharajan-Krishnakumar. 1987. Full-information estimations of a system of simultaneous equations with error component structure. *Econometric Theory* 3: 223–246.

Baltagi, B. H. 2001. *Econometric Analysis of Panel Data*. 2d ed. New York: John Wiley & Sons.

Baltagi, B. H. and Y. Chang. 1994. Incomplete panels: A comparative study of alternative estimators for the unbalanced one-way error component regression model. *Journal of Econometrics* 62: 67–89.

——. 2000. Simultaneous equations with incomplete panels. *Econometric Theory* 16: 269–279.

Baltagi, B. H. and Q. Li. 1992. A note on the estimation of simultaneous equations with error components. *Econometric Theory* 8: 113–119.

Swamy, P. A. V. B. and S. S. Arora. 1972. The exact finite sample properties of the estimators of coefficients in the error components regression models. *Econometrica* 40: 261–275.

White, H. 1984. *Asymptotic Theory for Econometricians*. New York: Academic Press.

Wooldridge, J. M. 1990. A note on the Lagrange multiple and F statistics for two-stage least squares regressions. *Economics Letters* 34: 151–155.

(Continued on next page)

Also See

Complementary:	[XT] **xtdata**, [XT] **xtdes**, [XT] **xtsum**, [XT] **xttab**,
	[R] **adjust**, [R] **lincom**, [R] **mfx**, [R] **nlcom**, [R] **predict**, [R] **predictnl**,
	[R] **test**, [R] **testnl**, [R] **vce**,
	[TS] **tsset**
Related:	[XT] **xtreg**, [XT] **xtregar**; [XT] **xtabond**, [XT] **xtgee**, [XT] **xthtaylor**,
	[XT] **xtintreg**, [XT] **xttobit**
Background:	[U] **16.5 Accessing coefficients and standard errors**,
	[U] **23 Estimation and post-estimation commands**,
	[XT] **xt**

Title

xtlogit — Fixed-effects, random-effects, and population-averaged logit models

Syntax

Random-effects model

> xtlogit *depvar* [*varlist*] [*weight*] [if *exp*] [in *range*] [, re i(*varname*) or
>
> quad(#) noconstant noskip level(#) offset(*varname*) nolog nodisplay
>
> *maximize_options*]

Conditional fixed-effects model

> xtlogit *depvar* [*varlist*] [*weight*] [if *exp*] [in *range*] , fe [i(*varname*) or
>
> noskip level(#) offset(*varname*) nolog nodisplay *maximize_options*]

Population-averaged model

> xtlogit *depvar* [*varlist*] [*weight*] [if *exp*] [in *range*] , pa [i(*varname*) or robust
>
> noconstant level(#) offset(*varname*) nolog nodisplay
>
> *xtgee_options maximize_options*]

by ... : may be used with xtlogit; see [R] **by**.

iweights, fweights, and pweights are allowed for the population-averaged model and iweights are allowed for the fixed-effects and random-effects models; see [U] **14.1.6 weight**. Note that weights must be constant within panels.

xtlogit shares the features of all estimation commands; see [U] **23 Estimation and post-estimation commands**.

Syntax for predict

Random-effects model

> predict [*type*] *newvarname* [if *exp*] [in *range*] [, [xb | pu0 | stdp] nooffset]

Fixed-effects model

> predict [*type*] *newvarname* [if *exp*] [in *range*] [, [p | xb | stdp] nooffset]

Population-averaged model

> predict [*type*] *newvarname* [if *exp*] [in *range*] [, [mu | rate | xb | stdp]
>
> nooffset]

These statistics are available both in and out of sample; type predict ... if e(sample) ... if wanted only for the estimation sample.

Note that the predicted probability for the fixed-effects model is conditional on there being only one outcome per group. See [R] **clogit** for details.

Description

xtlogit fits random-effects, conditional fixed-effects, and population-averaged logit models. Whenever we refer to a fixed-effects model, we mean the conditional fixed-effects model.

Note: xtlogit, re is slow since it is calculated by quadrature; see *Methods and Formulas*. Computation time is roughly proportional to the number of points used for the quadrature. The default is quad(12). Simulations indicate that increasing it does not appreciably change the estimates for the coefficients or their standard errors. See [XT] **quadchk**.

By default, the population-averaged model is an equal-correlation model; xtlogit assumes corr(exchangeable). See [XT] **xtgee** for details on how to fit other population-averaged models.

See [R] **logistic** for a list of related estimation commands.

Options

re requests the random-effects estimator. re is the default if none of re, fe, and pa are specified.

fe requests the fixed-effects estimator.

pa requests the population-averaged estimator.

i(*varname*) specifies the variable name that contains the unit to which the observation belongs. You can specify the i() option the first time you estimate, or you can use the iis command to set i() beforehand. Note that it is not necessary to specify i() if the data have been previously tsset, or if iis has been previously specified—in these cases, the group variable is taken from the previous setting. See [XT] **xt**.

or reports the estimated coefficients transformed to odds ratios; i.e., e^b rather than b. Standard errors and confidence intervals are similarly transformed. This option affects how results are displayed, not how they are estimated. or may be specified at estimation or when replaying previously estimated results.

quad(*#*) specifies the number of points to use in the quadrature approximation of the integral. The default is quad(12). See [XT] **quadchk**.

noconstant suppresses the constant term (intercept) in the model.

noskip specifies that a full maximum-likelihood model with only a constant for the regression equation be fitted. This model is not displayed, but is used as the base model to compute a likelihood-ratio test for the model test statistic displayed in the estimation header. By default, the overall model test statistic is an asymptotically equivalent Wald test of all the parameters in the regression equation being zero (except the constant). For many models, this option can substantially increase estimation time.

level(*#*) specifies the confidence level, in percent, for confidence intervals. The default is level(95) or as set by set level; see [U] **23.6 Specifying the width of confidence intervals**.

offset(*varname*) specifies that *varname* is to be included in the model with its coefficient constrained to be 1.

robust specifies that the Huber/White/sandwich estimator of variance is to be used in place of the IRLS variance estimator; see [XT] **xtgee**. This alternative produces valid standard errors even if the correlations within group are not as hypothesized by the specified correlation structure. It does, however, require that the model correctly specifies the mean. As such, the resulting standard errors are labeled "semi-robust" instead of "robust". Note that although there is no cluster() option, results are as if there were a cluster() option and you specified clustering on i().

`nolog` suppresses the iteration log.

`nodisplay` is for programmers. It suppresses the display of the header and the coefficients.

xtgee_options specifies any other options allowed by `xtgee` for `family(binomial) link(logit)` such as `corr()`; see [XT] **xtgee**.

maximize_options control the maximization process; see [R] **maximize**. Use the `trace` option to view parameter convergence. Use the `ltol(#)` option to relax the convergence criterion; the default is 1e−6 during specification searches.

Options for predict

`xb` calculates the linear prediction. This is the default for the random-effects model.

`p` calculates the predicted probability of a positive outcome conditional on one positive outcome within group. This is the default for the fixed-effects model.

`mu` and `rate` both calculate the predicted probability of *depvar*. `mu` takes into account the `offset()`, and `rate` ignores those adjustments. `mu` and `rate` are equivalent if you did not specify `offset()`. `mu` is the default for the population-averaged model.

`pu0` calculates the probability of a positive outcome, assuming that the random effect for that observation's panel is zero ($\nu = 0$). Note that this may not be similar to the proportion of observed outcomes in the group.

`stdp` calculates the standard error of the linear prediction.

`nooffset` is relevant only if you specified `offset(`*varname*`)` for `xtlogit`. It modifies the calculations made by `predict` so that they ignore the offset variable; the linear prediction is treated as $\mathbf{x}_{it}\boldsymbol{\beta}$ rather than $\mathbf{x}_{it}\boldsymbol{\beta} + \text{offset}_{it}$.

Remarks

`xtlogit` is a convenience command if you want the population-averaged model. Typing

        . xtlogit ..., pa ...

is equivalent to typing

        . xtgee ..., ... family(binomial) link(logit) corr(exchangeable)

It is also a convenience command if you want the fixed-effects model. Typing

        . xtlogit ..., fe i(*varname*) ...

is equivalent to typing

        . clogit ..., group(*varname*) ...

Thus, also see [XT] **xtgee** and [R] **clogit** for information about `xtlogit`.

By default, or when `re` is specified, `xtlogit` fits via maximum-likelihood the random-effects model

$$\Pr(y_{it} \neq 0 | \mathbf{x}_{it}) = P(\mathbf{x}_{it}\boldsymbol{\beta} + \nu_i)$$

for $i = 1, \ldots, n$ panels, $t = 1, \ldots, n_i$, ν_i are iid $N(0, \sigma_\nu^2)$, and $P(z) = \{1 + \exp(-z)\}^{-1}$.

Underlying this model is the variance components model

$$y_{it} \neq 0 \iff \mathbf{x}_{it}\boldsymbol{\beta} + \nu_i + \epsilon_{it} > 0$$

where ϵ_{it} are iid logistic distributed with mean zero and variance $\sigma_\epsilon^2 = \pi^2/3$, independently of ν_i.

▷ Example

You are studying unionization of women in the United States and are using the union dataset; see [XT] **xt**. You wish to fit a random-effects model of union membership:

```
. use http://www.stata-press.com/data/r8/union
(NLS Women 14-24 in 1968)

. xtlogit union age grade not_smsa south southXt, i(id) nolog

Random-effects logistic regression          Number of obs      =      26200
Group variable (i): idcode                  Number of groups   =       4434

Random effects u_i ~ Gaussian               Obs per group: min =          1
                                                           avg =        5.9
                                                           max =         12

                                            Wald chi2(5)       =     221.95
Log likelihood  = -10556.294                Prob > chi2        =     0.0000
```

union	Coef.	Std. Err.	z	P>|z|	[95% Conf. Interval]	
age	.0092401	.0044368	2.08	0.037	.0005441	.0179361
grade	.0840066	.0181622	4.63	0.000	.0484094	.1196038
not_smsa	-.2574574	.0844771	-3.05	0.002	-.4230294	-.0918854
south	-1.152854	.1108294	-10.40	0.000	-1.370075	-.9356323
southXt	.0237933	.0078548	3.03	0.002	.0083982	.0391884
_cons	-3.25016	.2622898	-12.39	0.000	-3.764238	-2.736081
/lnsig2u	1.669888	.0430016			1.585607	1.75417
sigma_u	2.304685	.0495526			2.209582	2.403882
rho	.6175213	.0101565			.5974278	.6372209

```
Likelihood-ratio test of rho=0: chibar2(01) =  5978.89 Prob >= chibar2 = 0.000
```

The output includes the additional panel-level variance component. This is parameterized as the log of the standard deviation $\ln(\sigma_\nu)$ (labeled `lnsig2u` in the output). The standard deviation σ_ν is also included in the output labeled `sigma_u` together with ρ (labeled `rho`),

$$\rho = \frac{\sigma_\nu^2}{\sigma_\nu^2 + \sigma_\epsilon^2}$$

which is the proportion of the total variance contributed by the panel-level variance component.

When `rho` is zero, the panel-level variance component is unimportant, and the panel estimator is not different from the pooled estimator. A likelihood-ratio test of this is included at the bottom of the output. This test formally compares the pooled estimator (logit) with the panel estimator.

As an alternative to the random-effects specification, you might want to fit an equal-correlation logit model:

(*Continued on next page*)

```
. xtlogit union age grade not_smsa south southXt, i(id) pa
```

```
Iteration 1: tolerance = .07495101
Iteration 2: tolerance = .00626455
Iteration 3: tolerance = .00030986
Iteration 4: tolerance = .00001432
Iteration 5: tolerance = 6.699e-07
```

GEE population-averaged model				Number of obs	=	26200
Group variable:			idcode	Number of groups	=	4434
Link:			logit	Obs per group: min =		1
Family:			binomial	avg =		5.9
Correlation:			exchangeable	max =		12
				Wald chi2(5)	=	233.60
Scale parameter:			1	Prob > chi2	=	0.0000

union	Coef.	Std. Err.	z	P>|z|	[95% Conf. Interval]	
age	.0053241	.0024988	2.13	0.033	.0004265	.0102216
grade	.0595076	.0108311	5.49	0.000	.0382791	.0807361
not_smsa	-.1224955	.0483137	-2.54	0.011	-.2171887	-.0278024
south	-.7270863	.0675522	-10.76	0.000	-.8594861	-.5946865
southXt	.0151984	.0045586	3.33	0.001	.0062638	.024133
_cons	-2.01111	.15439	-13.03	0.000	-2.313709	-1.708512

◁

▷ Example

xtlogit with the pa option allows a robust option, so we can obtain the population-averaged logit estimator with the robust variance calculation by typing

```
. xtlogit union age grade not_smsa south southXt, i(id) pa robust nolog
```

GEE population-averaged model				Number of obs	=	26200
Group variable:			idcode	Number of groups	=	4434
Link:			logit	Obs per group: min =		1
Family:			binomial	avg =		5.9
Correlation:			exchangeable	max =		12
				Wald chi2(5)	=	152.01
Scale parameter:			1	Prob > chi2	=	0.0000

(standard errors adjusted for clustering on idcode)

union	Coef.	Semi-robust Std. Err.	z	P>|z|	[95% Conf. Interval]	
age	.0053241	.0037494	1.42	0.156	-.0020246	.0126727
grade	.0595076	.0133482	4.46	0.000	.0333455	.0856697
not_smsa	-.1224955	.0613646	-2.00	0.046	-.2427678	-.0022232
south	-.7270863	.0870278	-8.35	0.000	-.8976577	-.5565149
southXt	.0151984	.006613	2.30	0.022	.0022371	.0281596
_cons	-2.01111	.2016405	-9.97	0.000	-2.406319	-1.615902

These standard errors are somewhat larger than those obtained without the robust option.

◁

Finally, we can also fit a fixed-effects model to these data (see also [R] **clogit** for details):

```
. xtlogit union age grade not_smsa south southXt, i(id) fe
note: multiple positive outcomes within groups encountered.
note: 2744 groups (14165 obs) dropped due to all positive or
      all negative outcomes.
Iteration 0:   log likelihood = -4541.9044
Iteration 1:   log likelihood = -4511.1353
Iteration 2:   log likelihood = -4511.1042
```

Conditional fixed-effects logistic regression	Number of obs	=	12035
Group variable (i): idcode	Number of groups	=	1690

```
                                    Obs per group: min =        2
                                                   avg =      7.1
                                                   max =       12

                                    LR chi2(5)          =    78.16
Log likelihood  = -4511.1042        Prob > chi2         =   0.0000
```

union	Coef.	Std. Err.	z	P>|z|	[95% Conf. Interval]	
age	.0079706	.0050283	1.59	0.113	-.0018848	.0178259
grade	.0811808	.0419137	1.94	0.053	-.0009686	.1633302
not_smsa	.0210368	.113154	0.19	0.853	-.2007411	.2428146
south	-1.007318	.1500491	-6.71	0.000	-1.301409	-.7132271
southXt	.0263495	.0083244	3.17	0.002	.010034	.0426649

Saved Results

xtlogit, re saves in e():

Scalars

e(N)	number of observations	e(g_avg)	average group size
e(N_g)	number of groups	e(chi2)	χ^2
e(df_m)	model degrees of freedom	e(chi2_c)	χ^2 for comparison test
e(ll)	log likelihood	e(rho)	ρ
e(ll_0)	log likelihood, constant-only model	e(sigma_u)	panel-level standard deviation
e(ll_c)	log likelihood, comparison model	e(N_cd)	number of completely determined obs.
e(g_max)	largest group size	e(n_quad)	number of quadrature points
e(g_min)	smallest group size		

Macros

e(cmd)	xtlogit	e(chi2type)	Wald or LR; type of model χ^2 test
e(depvar)	name of dependent variable	e(chi2_ct)	Wald or LR; type of model χ^2 test
e(title)	title in estimation output		corresponding to e(chi2_c)
e(ivar)	variable denoting groups	e(distrib)	Gaussian; the distribution of the
e(wtype)	weight type		random effect
e(wexp)	weight expression	e(crittype)	optimization criterion
e(offset)	offset	e(predict)	program used to implement predict

Matrices

e(b)	coefficient vector	e(V)	variance–covariance matrix of the
			estimators

Functions

e(sample)	marks estimation sample

`xtlogit, fe` saves in `e()`:

Scalars

`e(N)`	number of observations	`e(g_max)`	largest group size
`e(N_g)`	number of groups	`e(g_min)`	smallest group size
`e(df_m)`	model degrees of freedom	`e(g_avg)`	average group size
`e(ll)`	log likelihood	`e(chi2)`	χ^2
`e(ll_0)`	log likelihood, constant-only model		

Macros

`e(cmd)`	`clogit`	`e(wtype)`	weight type
`e(cmd2)`	`xtlogit`	`e(wexp)`	weight expression
`e(depvar)`	name of dependent variable	`e(chi2type)`	LR; type of model χ^2 test
`e(title)`	title in estimation output	`e(crittype)`	optimization criterion
`e(ivar)`	variable denoting groups	`e(predict)`	program used to implement `predict`
`e(offset)`	offset		

Matrices

`e(b)`	coefficient vector	`e(V)`	variance–covariance matrix of the estimators

Functions

`e(sample)`	marks estimation sample

`xtlogit, pa` saves in `e()`:

Scalars

`e(N)`	number of observations	`e(deviance)`	deviance
`e(N_g)`	number of groups	`e(chi2_dev)`	χ^2 test of deviance
`e(df_m)`	model degrees of freedom	`e(dispers)`	deviance dispersion
`e(g_max)`	largest group size	`e(chi2_dis)`	χ^2 test of deviance dispersion
`e(g_min)`	smallest group size	`e(tol)`	target tolerance
`e(g_avg)`	average group size	`e(dif)`	achieved tolerance
`e(chi2)`	χ^2	`e(phi)`	scale parameter
`e(df_pear)`	degrees of freedom for Pearson χ^2		

Macros

`e(cmd)`	`xtgee`	`e(scale)`	x2, dev, phi, or #; scale parameter
`e(cmd2)`	`xtlogit`	`e(ivar)`	variable denoting groups
`e(depvar)`	name of dependent variable	`e(vcetype)`	covariance estimation method
`e(family)`	binomial	`e(chi2type)`	Wald; type of model χ^2 test
`e(link)`	logit; link function	`e(offset)`	offset
`e(corr)`	correlation structure	`e(predict)`	program used to implement `predict`
`e(crittype)`	optimization criterion		

Matrices

`e(b)`	coefficient vector	`e(R)`	estimated working correlation matrix
`e(V)`	variance–covariance matrix of the estimators		

Functions

`e(sample)`	marks estimation sample

Methods and Formulas

`xtlogit` is implemented as an ado-file.

`xtlogit` reports the population-averaged results obtained by using `xtgee, family(binomial)` `link(logit)` to obtain estimates. The fixed-effects results are obtained using `clogit`. See [XT] **xtgee** and [R] **clogit** for details on the methods and formulas.

Assuming a normal distribution, $N(0, \sigma_\nu^2)$, for the random effects ν_i, we have that

$$\Pr(y_{i1}, \ldots, y_{in_i} | \mathbf{x}_{i1}, \ldots, \mathbf{x}_{in_i}) = \int_{-\infty}^{\infty} \frac{e^{-\nu_i^2/2\sigma_\nu^2}}{\sqrt{2\pi}\sigma_\nu} \left\{ \prod_{t=1}^{n_i} F(y_{it}, \mathbf{x}_{it}\boldsymbol{\beta} + \nu_i) \right\} d\nu_i$$

where

$$F(y, z) = \begin{cases} \dfrac{1}{1 + \exp(-z)} & \text{if } y \neq 0 \\ \dfrac{1}{1 + \exp(z)} & \text{otherwise} \end{cases}$$

and we can approximate the integral with M-point Gauss–Hermite quadrature

$$\int_{-\infty}^{\infty} e^{-x^2} g(x)dx \approx \sum_{m=1}^{M} w_m^* g(a_m^*)$$

where the w_m^* denote the quadrature weights and the a_m^* denote the quadrature abscissas. The log-likelihood L, where $\tau = \sigma_\nu^2/(\sigma_\nu^2 + 1)$, is then calculated using the quadrature

$$L = \sum_{i=1}^{n} w_i \log \left\{ \Pr(y_{i1}, \ldots, y_{in_i} | \mathbf{x}_{i1}, \ldots, \mathbf{x}_{in_i}) \right\}$$

$$\approx \sum_{i=1}^{n} w_i \log \left[\frac{1}{\sqrt{\pi}} \sum_{m=1}^{M} w_m^* \prod_{t=1}^{n_i} F \left\{ y_{it}, \mathbf{x}_{it}\boldsymbol{\beta} + a_m^* \left(\frac{2\tau}{1 - \tau} \right)^{1/2} \right\} \right]$$

where w_i is the user-specified weight for panel i; if no weights are specified, $w_i = 1$.

The quadrature formula requires that the integrated function be well-approximated by a polynomial. As the number of time periods becomes large (as panel size gets large),

$$\prod_{t=1}^{n_i} F(y_{it}, \mathbf{x}_{it}\boldsymbol{\beta} + \nu_i)$$

is no longer well-approximated by a polynomial. As a general rule of thumb, you should use this quadrature approach only for small to moderate panel sizes (based on simulations, 50 is a reasonably safe upper bound). However, if the data really come from random-effects logit and τ is not too large (less than, say, .3), then the panel size could be 500 and the quadrature approximation would still be fine. If the data are not random-effects logit or τ is large (bigger than, say, .7), then the quadrature approximation may be poor for panel sizes larger than 10. The `quadchk` command should be used to investigate the applicability of the numeric technique used in this command.

References

Conway, M. R. 1990. A random effects model for binary data. *Biometrics* 46: 317–328.

Liang, K.-Y. and S. L. Zeger. 1986. Longitudinal data analysis using generalized linear models. *Biometrika* 73: 13–22.

Neuhaus, J. M. 1992. Statistical methods for longitudinal and clustered designs with binary responses. *Statistical Methods in Medical Research* 1: 249–273.

Neuhaus, J. M., J. D. Kalbfleisch, and W. W. Hauck. 1991. A comparison of cluster-specific and population-averaged approaches for analyzing correlated binary data. *International Statistical Review* 59: 25–35.

Pendergast, J. F., S. J. Gange, M. A. Newton, M. J. Lindstrom, M. Palta, and M. R. Fisher. 1996. A survey of methods for analyzing clustered binary response data. *International Statistical Review* 64: 89–118.

Also See

Complementary:	[XT] **quadchk**, [XT] **xtdata**, [XT] **xtdes**, [XT] **xtsum**, [XT] **xttab**, [R] **adjust**, [R] **lincom**, [R] **mfx**, [R] **nlcom**, [R] **predict**, [R] **predictnl**, [R] **test**, [R] **testnl**, [R] **vce**
Related:	[XT] **xtcloglog**, [XT] **xtgee**, [XT] **xtprobit**, [R] **clogit**, [R] **logit**
Background:	[U] **16.5 Accessing coefficients and standard errors**, [U] **23 Estimation and post-estimation commands**, [U] **23.14 Obtaining robust variance estimates**, [XT] **xt**

Title

> **xtnbreg** — Fixed-effects, random-effects, & population-averaged negative binomial models

Syntax

Random-effects and conditional fixed-effects overdispersion models

> xtnbreg *depvar* [*varlist*] [*weight*] [if *exp*] [in *range*] [, [re | fe] i(*varname*)
>
> irr noconstant constraints(*numlist*) noskip exposure(*varname*)
>
> offset(*varname*) level(#) nolog *maximize_options*]

Population-averaged model

> xtnbreg *depvar* [*varlist*] [*weight*] [if *exp*] [in *range*] , pa [i(*varname*)
>
> irr robust noconstant exposure(*varname*) offset(*varname*) level(#)
>
> nolog *xtgee_options* *maximize_options*]

by ... : may be used with xtnbreg; see [R] **by**.

iweights, fweights, and pweights are allowed for the population-averaged model and iweights are allowed in the random-effects and fixed-effects models; see [U] **14.1.6 weight**. Note that weights must be constant within panels.

xtnbreg shares the features of all estimation commands; see [U] **23 Estimation and post-estimation commands**.

Syntax for predict

Random-effects and conditional fixed-effects overdispersion models

> predict [*type*] *newvarname* [if *exp*] [in *range*] [, [xb | stdp | nu0 | iru0]
>
> nooffset]

Population-averaged model

> predict [*type*] *newvarname* [if *exp*] [in *range*] [, [mu | rate | xb | stdp]
>
> nooffset]

These statistics are available both in and out of sample; type predict ... if e(sample) ... if wanted only for the estimation sample.

(Continued on next page)

Description

xtnbreg fits random-effects overdispersion models, conditional fixed-effects overdispersion models, and population-averaged negative binomial models. Here, "random-effects" and "fixed-effects" apply to the distribution of the dispersion parameter, and not to the $x\beta$ term in the model. In the random-effects and fixed-effects overdispersion models, the dispersion is the same for all elements in the same group (i.e., elements with the same value of the i() variable). In the random-effects model, the dispersion varies randomly from group to group such that the inverse of one plus the dispersion follows a Beta(r, s) distribution. In the fixed-effects model, the dispersion parameter in a group can take on any value, since a conditional likelihood is used in which the dispersion parameter drops out of the estimation.

By default, the population-averaged model is an equal-correlation model; xtnbreg assumes corr(exchangeable). See [XT] **xtgee** for details on this option to fit other population-averaged models.

Options

re requests the random-effects estimator. re is the default if none of re, fe, and pa are specified.

fe requests the conditional fixed-effects estimator.

pa requests the population-averaged estimator.

i(*varname*) specifies the variable name that contains the unit to which the observation belongs. You can specify the i() option the first time you estimate, or you can use the iis command to set i() beforehand. Note that it is not necessary to specify i() if the data have been previously tsset, or if iis has been previously specified—in these cases, the time variable is taken from the previous setting. See [XT] **xt**.

irr reports exponentiated coefficients e^b rather than coefficients b. For the negative binomial model, exponentiated coefficients have the interpretation of incidence rate ratios.

noconstant suppresses the constant term (intercept) in the model.

constraints(*numlist*) specifies by number the linear constraints to be applied during estimation. The default is to perform unconstrained estimation. Constraints are specified using the constraint command; see [R] **constraint**. constraints(*numlist*) may not be specified with pa.

noskip specifies that a full maximum-likelihood model with only a constant for the regression equation be fitted. This constant-only model is used as the base model to compute a likelihood-ratio χ^2 statistic for the model test. By default, the model test uses an asymptotically equivalent Wald χ^2 statistic. For many models, this option can substantially increase estimation time.

exposure(*varname*) and offset(*varname*) are different ways of specifying the same thing. exposure() specifies a variable that reflects the amount of exposure over which the *depvar* events were observed for each observation; ln(*varname*) with its coefficient constrained to be 1 is entered into the regression equation. offset() specifies a variable that is to be entered directly into the regression equation with its coefficient constrained to be 1; thus, exposure is assumed to be $e^{varname}$.

level(*#*) specifies the confidence level, in percent, for confidence intervals. The default is level(95) or as set by set level; see [U] **23.6 Specifying the width of confidence intervals**.

robust (pa only) specifies that the Huber/White/sandwich estimator of variance is to be used in place of the IRLS variance estimator; see [XT] **xtgee**. This alternative produces valid standard errors even if the correlations within group are not as hypothesized by the specified correlation structure. It does, however, require that the model correctly specifies the mean. As such, the resulting standard errors are labeled "semi-robust" instead of "robust". Note that although there is no cluster() option, results are as if there were a cluster() option and you specified clustering on i().

nolog suppresses the iteration log.

xtgee_options specifies any other options allowed by xtgee for family(nbinom) link(log); see [XT] **xtgee**.

maximize_options control the maximization process; see [R] **maximize**. Use the trace option to view parameter convergence.

Options for predict

xb calculates the linear prediction. This is the default for the random-effects and fixed-effects models.

stdp calculates the standard error of the linear prediction.

nu0 calculates the predicted number of events, assuming a zero fixed/random effect.

iru0 calculates the predicted incidence rate, assuming a zero fixed/random effect.

mu and rate both calculate the predicted probability of *depvar*. mu takes into account the offset(), and rate ignores those adjustments. mu and rate are equivalent if you did not specify offset(). mu is the default for the population-averaged model.

nooffset is relevant only if you specified offset(*varname*) for xtnbreg. It modifies the calculations made by predict so that they ignore the offset variable; the linear prediction is treated as $\mathbf{x}_{it}\boldsymbol{\beta}$ rather than $\mathbf{x}_{it}\boldsymbol{\beta} + \text{offset}_{it}$.

Remarks

xtnbreg is a convenience command if you want the population-averaged model. Typing

```
. xtnbreg ..., ... pa exposure(time)
```

is equivalent to typing

```
. xtgee ..., ... family(nbinom) link(log) corr(exchangeable) exposure(time)
```

Thus, also see [XT] **xtgee** for information about xtnbreg.

By default, or when re is specified, xtnbreg fits a maximum-likelihood random-effects overdispersion model.

▷ Example

You have (fictional) data on injury "incidents" incurred among 20 airlines in each of 4 years. (Incidents range from major injuries to exceedingly minor ones.) The government agency in charge of regulating airlines has run an experimental safety training program, and, in each of the years, some airlines have participated and some have not. You now wish to analyze whether the "incident" rate is affected by the program. You choose to estimate using random-effects negative binomial regression because the dispersion might vary across the airlines because of unidentified airline-specific reasons. Your measure of exposure is passenger miles for each airline in each year.

```
. use http://www.stata-press.com/data/r8/airacc

. xtnbreg i_cnt inprog, i(airline) exposure(pmiles) irr nolog
```

```
Random-effects negative binomial regression    Number of obs      =        80
Group variable (i): airline                    Number of groups   =        20

Random effects u_i ~ Beta                       Obs per group: min =         4
                                                               avg =       4.0
                                                               max =         4

                                               Wald chi2(1)       =      2.04
Log likelihood  = -265.38202                   Prob > chi2        =    0.1532
```

i_cnt	IRR	Std. Err.	z	P>|z|	[95% Conf. Interval]	
inprog	.911673	.0590277	-1.43	0.153	.8030206	1.035027
pmiles	(exposure)					
/ln_r	4.794991	.951781			2.929535	6.660448
/ln_s	3.268052	.4709033			2.345098	4.191005
r	120.9033	115.0735			18.71892	780.9007
s	26.26013	12.36598			10.4343	66.08918

```
Likelihood-ratio test vs. pooled: chibar2(01) =    19.03 Prob>=chibar2 = 0.000
```

In the output above, the /ln_r and /ln_s lines refer to $\ln(r)$ and $\ln(s)$, where the inverse of one plus the dispersion is assumed to follow a Beta(r, s) distribution. The output also includes a likelihood-ratio test, which compares the panel estimator with the pooled estimator (i.e., a negative binomial estimator with constant dispersion).

You find that the incidence rate for accidents is not significantly different for participation in the program, and that the panel estimator is significantly different from the pooled estimator.

We may alternatively fit a fixed-effects overdispersion model:

```
. xtnbreg i_cnt inprog, i(airline) exposure(pmiles) irr fe nolog
```

```
Conditional FE negative binomial regression    Number of obs      =        80
Group variable (i): airline                    Number of groups   =        20

                                                Obs per group: min =         4
                                                               avg =       4.0
                                                               max =         4

                                               Wald chi2(1)       =      2.11
Log likelihood  = -174.25143                   Prob > chi2        =    0.1463
```

i_cnt	IRR	Std. Err.	z	P>|z|	[95% Conf. Interval]	
inprog	.9062669	.0613917	-1.45	0.146	.793587	1.034946
pmiles	(exposure)					

◁

(Continued on next page)

▷ Example

We rerun our previous example, but this time we fit a robust equal-correlation population-averaged model:

```
. xtnbreg i_cnt inprog, i(airline) exposure(pmiles) eform robust pa nolog
GEE population-averaged model          Number of obs      =        80
Group variable:                airline  Number of groups   =        20
Link:                              log   Obs per group: min =         4
Family:         negative binomial(k=1)                 avg =       4.0
Correlation:              exchangeable                 max =         4
                                        Wald chi2(1)       =      1.28
Scale parameter:                     1   Prob > chi2        =    0.2571
               (standard errors adjusted for clustering on airline)
```

i_cnt	IRR	Semi-robust Std. Err.	z	P>|z|	[95% Conf. Interval]
inprog	.927275	.0617857	-1.13	0.257	.8137513 1.056636
pmiles	(exposure)				

We may compare this with a pooled estimator with clustered robust-variance estimates:

```
. nbreg i_cnt inprog, exposure(pmiles) robust cluster(airline) irr nolog
Negative binomial regression            Number of obs      =        80
                                        Wald chi2(1)       =      0.60
Log likelihood = -274.55077             Prob > chi2        =    0.4369
               (standard errors adjusted for clustering on airline)
```

i_cnt	IRR	Robust Std. Err.	z	P>|z|	[95% Conf. Interval]
inprog	.9429015	.0713091	-0.78	0.437	.8130032 1.093555
pmiles	(exposure)				
/lnalpha	-2.835089	.3351784			-3.492027 -2.178151
alpha	.0587133	.0196794			.0304391 .1132507

◁

(*Continued on next page*)

Saved Results

xtnbreg, re saves in e():

Scalars

e(N)	number of observations	e(ll_c)	log likelihood, comparison model
e(k)	number of estimated parameters	e(df_m)	model degrees of freedom
e(k_eq)	number of equations	e(chi2)	model χ^2
e(k_dv)	number of dependent variables	e(p)	model significance
e(N_g)	number of groups	e(chi2_c)	χ^2 for comparison test
e(g_min)	smallest group size	e(r)	value of r in Beta(r,s)
e(g_avg)	average group size	e(s)	value of s in Beta(r,s)
e(g_max)	largest group size	e(ic)	number of iterations
e(ll)	log likelihood	e(rc)	return code
e(ll_0)	log likelihood, constant-only model		

Macros

e(cmd)	xtnbreg	e(opt)	type of optimization
e(cmd2)	xtn_re	e(chi2type)	Wald or LR; type of model χ^2 test
e(depvar)	name of dependent variable	e(chi2_ct)	Wald or LR; type of model χ^2 test
e(title)	title in estimation output		corresponding to e(chi2_c)
e(ivar)	variable denoting groups	e(offset)	offset
e(wtype)	weight type	e(distrib)	Beta; the distribution of the
e(wexp)	weight expression		random effect
e(method)	estimation method	e(crittype)	optimization criterion
e(user)	name of likelihood-evaluation program	e(predict)	program used to implement predict

Matrices

e(b)	coefficient vector	e(V)	variance–covariance matrix of the estimators

Functions

e(sample)	marks estimation sample

(Continued on next page)

`xtnbreg, fe` saves in `e()`:

Scalars

`e(N)`	number of observations	`e(ll)`	log likelihood
`e(k)`	number of estimated parameters	`e(ll_0)`	log likelihood, constant-only model
`e(k_eq)`	number of equations	`e(df_m)`	model degrees of freedom
`e(k_dv)`	number of dependent variables	`e(chi2)`	model χ^2
`e(N_g)`	number of groups	`e(p)`	model significance
`e(g_min)`	smallest group size	`e(ic)`	number of iterations
`e(g_avg)`	average group size	`e(rc)`	return code
`e(g_max)`	largest group size		

Macros

`e(cmd)`	xtnbreg	`e(method)`	requested estimation method
`e(cmd2)`	xtn_fe	`e(user)`	name of likelihood-evaluator program
`e(depvar)`	name of dependent variable	`e(opt)`	type of optimization
`e(title)`	title in estimation output	`e(chi2type)`	Wald or LR; type of model χ^2 test
`e(ivar)`	variable denoting groups	`e(offset)`	offset
`e(wtype)`	weight type	`e(crittype)`	optimization criterion
`e(wexp)`	weight expression	`e(predict)`	program used to implement predict

Matrices

`e(b)`	coefficient vector	`e(V)`	variance–covariance matrix of the estimators

Functions

`e(sample)`	marks estimation sample

`xtnbreg, pa` saves in `e()`:

Scalars

`e(N)`	number of observations	`e(deviance)`	deviance
`e(N_g)`	number of groups	`e(chi2_dev)`	χ^2 test of deviance
`e(g_min)`	smallest group size	`e(dispers)`	deviance dispersion
`e(g_avg)`	average group size	`e(chi2_dis)`	χ^2 test of deviance dispersion
`e(g_max)`	largest group size	`e(tol)`	target tolerance
`e(df_m)`	model degrees of freedom	`e(dif)`	achieved tolerance
`e(chi2)`	model χ^2	`e(phi)`	scale parameter
`e(df_pear)`	degrees of freedom for Pearson χ^2		

Macros

`e(cmd)`	xtgee	`e(ivar)`	variable denoting groups
`e(cmd2)`	xtnbreg	`e(vcetype)`	covariance estimation method
`e(depvar)`	name of dependent variable	`e(chi2type)`	Wald; type of model χ^2 test
`e(family)`	negative binomial(k=1)	`e(offset)`	offset
`e(link)`	log; link function	`e(nbalpha)`	α
`e(corr)`	correlation structure	`e(crittype)`	optimization criterion
`e(scale)`	x2, dev, phi, or #; scale parameter	`e(predict)`	program used to implement predict

Matrices

`e(b)`	coefficient vector	`e(V)`	variance–covariance matrix of the estimators
`e(R)`	estimated working correlation matrix		

Functions

`e(sample)`	marks estimation sample

Methods and Formulas

xtnbreg is implemented as an ado-file.

xtnbreg, pa reports the population-averaged results obtained by using xtgee, family(nbreg) link(log) to obtain estimates. See [XT] **xtgee** for details on the methods and formulas.

For the random-effects and fixed-effects overdispersion models, we let y_{it} be the count for the tth observation in the ith group. We begin with the model $y_{it} \mid \gamma_{it} \sim \text{Poisson}(\gamma_{it})$, where $\gamma_{it} \mid \delta_i \sim \text{gamma}(\lambda_{it}, 1/\delta_i)$ with $\lambda_{it} = \exp(\mathbf{x}_{it}\boldsymbol{\beta} + \text{offset}_{it})$ and δ_i is the dispersion parameter. This yields the model

$$\Pr(Y_{it} = y_{it} \mid \mathbf{x}_{it}, \delta_i) = \frac{\Gamma(\lambda_{it} + y_{it})}{\Gamma(\lambda_{it})\Gamma(y_{it} + 1)} \left(\frac{1}{1 + \delta_i}\right)^{\lambda_{it}} \left(\frac{\delta_i}{1 + \delta_i}\right)^{y_{it}}$$

Looking at within-group effects only, this specification yields a negative binomial model for the ith group with dispersion (variance divided by the mean) equal to $1 + \delta_i$; i.e., constant dispersion within group. Note that this parameterization of the negative binomial model differs from the default parameterization of nbreg, which has dispersion equal to $1 + \alpha \exp(\mathbf{x}\boldsymbol{\beta} + \text{offset})$; see [R] **nbreg**.

For a random-effects overdispersion model, we allow δ_i to vary randomly across groups; namely, we assume that $1/(1 + \delta_i) \sim \text{Beta}(r, s)$. The joint probability of the counts for the ith group is

$$\Pr(Y_{i1} = y_{i1}, \ldots, Y_{in_i} = y_{in_i} \mid \mathbf{X}_i) = \int \prod_{t=1}^{n_i} \Pr(Y_{it} = y_{it} \mid \mathbf{x}_{it}, \delta_i) \, f(\delta_i) \, d\delta_i$$

$$= \frac{\Gamma(r + s)\Gamma(r + \sum_{t=1}^{n_i} \lambda_{it})\Gamma(s + \sum_{t=1}^{n_i} y_{it})}{\Gamma(r)\Gamma(s)\Gamma(r + s + \sum_{t=1}^{n_i} \lambda_{it} + \sum_{t=1}^{n_i} y_{it})} \prod_{t=1}^{n_i} \frac{\Gamma(\lambda_{it} + y_{it})}{\Gamma(\lambda_{it})\Gamma(y_{it} + 1)}$$

for $\mathbf{X}_i = (\mathbf{x}_{i1}, \ldots, \mathbf{x}_{in_i})$. The resulting log likelihood is

$$\ln L = \sum_{i=1}^{n} w_i \left[\ln \Gamma(r + s) + \ln \Gamma\left(r + \sum_{k=1}^{n_i} \lambda_{ik}\right) + \ln \Gamma\left(s + \sum_{k=1}^{n_i} y_{ik}\right) - \ln \Gamma(r) - \ln \Gamma(s) \right.$$

$$\left. - \ln \Gamma\left(r + s + \sum_{k=1}^{n_i} \lambda_{ik} + \sum_{k=1}^{n_i} y_{ik}\right) + \sum_{t=1}^{n_i} \left\{ \ln \Gamma(\lambda_{it} + y_{it}) - \ln \Gamma(\lambda_{it}) - \ln \Gamma(y_{it} + 1) \right\} \right]$$

where $\lambda_{it} = \exp(\mathbf{x}_{it}\boldsymbol{\beta} + \text{offset}_{it})$ and w_i is the weight for the ith group.

For the fixed-effects overdispersion model, we condition the joint probability of the counts for each group on the sum of the counts for the group (i.e., the observed $\sum_{t=1}^{n_i} y_{it}$). This yields

$$\Pr(Y_{i1} = y_{i1}, \ldots, Y_{in_i} = y_{in_i} \mid \mathbf{X}_i, \textstyle\sum_{t=1}^{n_i} Y_{it} = \sum_{t=1}^{n_i} y_{it})$$

$$= \frac{\Gamma(\sum_{t=1}^{n_i} \lambda_{it})\Gamma(\sum_{t=1}^{n_i} y_{it} + 1)}{\Gamma(\sum_{t=1}^{n_i} \lambda_{it} + \sum_{t=1}^{n_i} y_{it})} \prod_{t=1}^{n_i} \frac{\Gamma(\lambda_{it} + y_{it})}{\Gamma(\lambda_{it})\Gamma(y_{it} + 1)}$$

The conditional log likelihood is

$$\ln L = \sum_{i=1}^{n} w_i \left[\ln \Gamma \left(\sum_{t=1}^{n_i} \lambda_{it} \right) + \ln \Gamma \left(\sum_{t=1}^{n_i} y_{it} + 1 \right) - \ln \Gamma \left(\sum_{t=1}^{n_i} \lambda_{it} + \sum_{t=1}^{n_i} y_{it} \right) \right.$$
$$\left. + \sum_{t=1}^{n_i} \left\{ \ln \Gamma(\lambda_{it} + y_{it}) - \ln \Gamma(\lambda_{it}) - \ln \Gamma(y_{it} + 1) \right\} \right]$$

See Hausman et al. (1984) for a more thorough development of the random-effects and fixed-effects models. Note that Hausman et al. (1984) use a δ that is the inverse of the δ we have used here. Also, see Cameron and Trivedi (1998) for a good textbook treatment of this model.

References

Cameron, A. C. and P. K. Trivedi. 1998. *Regression Analysis of Count Data.* New York: Cambridge University Press.

Hausman, J., B. H. Hall, and Z. Griliches. 1984. Econometric models for count data with an application to the patents–R & D relationship. *Econometrica* 52: 909–938.

Liang, K.-Y. and S. L. Zeger. 1986. Longitudinal data analysis using generalized linear models. *Biometrika* 73: 13–22.

Also See

Complementary:	[XT] **xtdata**, [XT] **xtdes**, [XT] **xtsum**, [XT] **xttab**,
	[R] **adjust**, [R] **constraint**, [R] **lincom**, [R] **mfx**, [R] **nlcom**,
	[R] **predict**, [R] **predictnl**, [R] **test**, [R] **testnl**, [R] **vce**
Related:	[XT] **xtgee**, [XT] **xtpoisson**,
	[R] **nbreg**
Background:	[U] **16.5 Accessing coefficients and standard errors**,
	[U] **23 Estimation and post-estimation commands**,
	[U] **23.14 Obtaining robust variance estimates**,
	[XT] **xt**

Title

> **xtpcse** — OLS or Prais–Winsten models with panel-corrected standard errors

Syntax

> **xtpcse** *depvar* [*varlist*] [*weight*] [**if** *exp*] [**in** *range*] [, <u>c</u>orrelation(*corr*) <u>het</u>only
>
> <u>i</u>ndependent <u>noc</u>onstant <u>casewise pairwise rho</u>type(*calc*) nmk np1 <u>detail</u>
>
> <u>l</u>evel(*#*)]

where *corr* is one of <u>i</u>ndependent | <u>ar</u>1 | <u>p</u>sar1

and *calc* is one of <u>reg</u>ress | freg | <u>tsc</u>orr | dw

by ... : may be used with xtpcse; see [R] **by**.

xtpcse is for use with time-series data; see [TS] **tsset**. You must tsset your data before using xtpcse.

depvar and *varlist* may contain time-series operators; see [U] **14.4.3 Time-series varlists**.

iweights and aweights are allowed; see [U] **14.1.6 weight**.

xtpcse shares the features of all estimation commands; see [U] **23 Estimation and post-estimation commands**.

Syntax for predict

> **predict** [*type*] *newvarname* [**if** *exp*] [**in** *range*] [, [xb | stdp]]

These statistics are available both in and out of sample; type predict ... if e(sample) ... if wanted only for the estimation sample.

Description

xtpcse calculates panel-corrected standard error (PCSE) estimates for linear cross-sectional time-series models where the parameters are estimated by either OLS or Prais–Winsten regression. When computing the standard errors and the variance–covariance estimates, xtpcse assumes that the disturbances are, by default, heteroskedastic and contemporaneously correlated across panels.

See [XT] **xtgls** for the generalized least squares estimator for these models.

Options

correlation(*corr*) specifies the form of assumed autocorrelation within panels.

correlation(independent), the default, specifies that there is no autocorrelation.

correlation(ar1) specifies that within panels, there is first-order autocorrelation AR(1), and that the coefficient of the AR(1) process is common to all the panels.

correlation(psar1) specifies that within panels, there is first-order autocorrelation, and that the coefficient of the AR(1) process is specific to each panel. psar1 stands for panel-specific AR(1).

150

`hetonly` and `independent` specify alternate forms for the assumed covariance of the disturbances across the panels. If neither is specified, the default is to assume that the disturbances are heteroskedastic (each panel has its own variance) and that the disturbances are contemporaneously correlated across the panels (each pair of panels has their own covariance). This is the standard model for PCSE.

`hetonly` specifies that the disturbances are assumed to be panel-level heteroskedastic only and that there is no assumed contemporaneous correlation across panels.

`independent` specifies that the disturbances are assumed to be independent across panels; that is, there is a single disturbance variance common to all observations.

`noconstant` suppresses the constant term (intercept) in the model.

`casewise` and `pairwise` specify how missing observations in unbalanced panels are to be treated when estimating the interpanel covariance matrix of the disturbances. The default is `casewise` selection.

`casewise` specifies that the entire covariance matrix is computed only on the observations (time periods) that are available for all panels. If an observation has missing data, all observations of that time period are excluded when estimating the covariance matrix of disturbances. Specifying `casewise` ensures that the estimated covariance matrix will be of full rank and will be positive definite.

`pairwise` specifies that for each element in the covariance matrix, all available observations (time periods) that are common to the two panels contributing to the covariance be used to compute the covariance.

Options `casewise` and `pairwise` have an effect only when the panels are unbalanced and neither `hetonly` nor `independent` is specified.

`rhotype(`*calc*`)` specifies the method to be used to calculate the autocorrelation parameter. Allowed strings for *calc* are

`regress`	regression using lags (the default)
`freg`	regression using leads
`dw`	Durbin–Watson calculation
`tscorr`	time series autocorrelation calculation

All the calculations are asymptotically equivalent and all are consistent; this is a rarely used option.

`nmk` specifies standard errors are to be normalized by $N - k$, where k is the number of parameters estimated, rather than N, the number of observations. Different authors have used one or the other normalization. Greene (2003, 322) recommends N, and notes that whether you use N or $N - k$ does not make the variance calculation unbiased in these models.

`np1` specifies that the panel-specific autocorrelations be weighted by T_i rather than the default $T_i - 1$ when estimating a common ρ for all panels, where T_i is the number of observations in panel i. This option has an effect only when panels are unbalanced and option `correlation(ar1)` is specified.

`detail` specifies that a detailed list of any gaps in the series be reported.

`level(`#`)` specifies the confidence level, in percent, for confidence intervals. The default is `level(95)` or as set by `set level`; see [U] **23.6 Specifying the width of confidence intervals**.

Options for predict

xb, the default, calculates the linear prediction.

stdp calculates the standard error of the linear prediction.

Remarks

xtpcse is an alternative to feasible generalized least squares (FGLS)—see [XT] **xtgls**—for fitting linear cross-sectional time-series models when the disturbances are not assumed to be independent and identically distributed (i.i.d.). Instead, the disturbances are assumed to be either heteroskedastic across panels or heteroskedastic and contemporaneously correlated across panels. The disturbances may also be assumed to be autocorrelated within panel, and the autocorrelation parameter may be either constant across panels or different for each panel.

We can write such models as

$$y_{it} = \mathbf{x}_{it}\boldsymbol{\beta} + \epsilon_{it}$$

where $i = 1, \dots, m$ is the number of units (or panels); $t = 1, \dots, T_i$; T_i is the number of time periods in panel i; and ϵ_{it} is a disturbance that may be autocorrelated along t or contemporaneously correlated across i.

This model can also be written panel-by-panel as

$$\begin{bmatrix} \mathbf{y}_1 \\ \mathbf{y}_2 \\ \vdots \\ \mathbf{y}_m \end{bmatrix} = \begin{bmatrix} \mathbf{X}_1 \\ \mathbf{X}_2 \\ \vdots \\ \mathbf{X}_m \end{bmatrix} \boldsymbol{\beta} + \begin{bmatrix} \epsilon_1 \\ \epsilon_2 \\ \vdots \\ \epsilon_m \end{bmatrix}$$

For a model with heteroskedastic disturbances and contemporaneous correlation but with no autocorrelation, the disturbance covariance matrix is assumed to be

$$E[\epsilon\epsilon'] = \boldsymbol{\Omega} = \begin{bmatrix} \sigma_{11}\mathbf{I}_{11} & \sigma_{12}\mathbf{I}_{12} & \cdots & \sigma_{1m}\mathbf{I}_{1m} \\ \sigma_{21}\mathbf{I}_{21} & \sigma_{22}\mathbf{I}_{22} & \cdots & \sigma_{2m}\mathbf{I}_{2m} \\ \vdots & \vdots & \ddots & \vdots \\ \sigma_{m1}\mathbf{I}_{m1} & \sigma_{m2}\mathbf{I}_{m2} & \cdots & \sigma_{mm}\mathbf{I}_{mm} \end{bmatrix}$$

where σ_{ii} is the variance of the disturbances for panel i, σ_{ij} is the covariance of the disturbances between panel i and panel j when the panels' time periods are matched, and $\mathbf{I}$ is a T_i by T_i identity matrix with balanced panels. Note that the panels need not be balanced for xtpcse, but the expression for the covariance of the disturbances will be more general if they are unbalanced.

This could also be written as

$$E(\epsilon'\epsilon) = \boldsymbol{\Sigma}_{m \times m} \otimes \mathbf{I}_{T_i \times T_i}$$

where $\boldsymbol{\Sigma}$ is the panel-by-panel covariance matrix and $\mathbf{I}$ is an identity matrix.

See [XT] **xtgls** for a full taxonomy and description of possible disturbance covariance structures.

xtpcse and xtgls follow two different estimation schemes for this family of models. xtpcse produces OLS estimates of the parameters when no autocorrelation is specified, or Prais–Winsten (see [TS] **prais**) estimates when autocorrelation is specified. If autocorrelation is specified, the estimates of the parameters are conditional on the estimates of the autocorrelation parameter(s). The estimate of the variance–covariance matrix of the parameters is an asymptotically efficient estimate under the assumed covariance structure of the disturbances, and uses the FGLS estimate of the disturbance covariance matrix; see Kmenta (1997, 121).

xtgls produces full FGLS parameter and variance–covariance estimates. These estimates are conditional on the estimates of the disturbance covariance matrix and are conditional on any autocorrelation parameters that are estimated; see Kmenta (1997), Greene (2003), Davidson and MacKinnon (1993), or Judge et al. (1985).

Both estimators are consistent as long as the conditional mean ($x_{it}\beta$) is correctly specified. If the assumed covariance structure is correct, FGLS estimates produced by xtgls are more efficient. Beck and Katz (1995) have shown, however, that the full FGLS variance–covariance estimates are typically unacceptably optimistic (anti-conservative) when used with the type of data analyzed by most social scientists—10 to 20 panels and 10 to 40 time periods per panel. They show that the OLS or Prais–Winsten estimates with PCSEs have coverage probabilities that are closer to nominal.

Since the covariance matrix elements, σ_{ij}, are estimated from panels i and j using those observations that have common time periods, estimators for this model achieve their asymptotic behavior as the T_i's approach infinity. This is in contrast to the random- and fixed-effects estimators that assume a different model and are asymptotic in the number of panels m; see [XT] **xtreg** for details of the random- and fixed-effects estimators.

Although xtpcse will allow other disturbance covariance structures, the term PCSE, as used in the literature, refers specifically to models that are both heteroskedastic and contemporaneously correlated across panels, with or without autocorrelation.

▷ Example

Grunfeld and Griliches (1960) performed an analysis of a company's current year gross investment (invest) as determined by the company's prior year market value (mvalue) and the prior year's value of the company's plant and equipment (kstock). The dataset includes 5 companies over 20 years, from 1935 through 1954, and is a classic dataset for demonstrating cross-sectional time-series analysis. Greene (2003, 329 and *http://www.prenhall.com/greene*) reproduces the dataset.

To use xtpcse, the data must be organized in "long form"; that is, each observation must represent a record for a specific company at a specific time; see [R] **reshape**. In the Grunfeld data, company is a categorical variable identifying the company and year is a variable recording the year. Here are the first few records:

```
. use http://www.stata-press.com/data/r8/grunfeld
. list in 1/5
```

	company	year	invest	mvalue	kstock	time
1.	1	1935	317.6	3078.5	2.8	1
2.	1	1936	391.8	4661.7	52.6	2
3.	1	1937	410.6	5387.1	156.9	3
4.	1	1938	257.7	2792.2	209.2	4
5.	1	1939	330.8	4313.2	203.4	5

To compute PCSEs, Stata must be able to identify the panel to which each observation belongs and be able to match the time periods across the panels. We tell Stata how to do this matching by specifying the time and panel variables using tsset; see [TS] **tsset**. Since the data are annual, we specify the yearly option.

```
. tsset company year, yearly
      panel variable:  company, 1 to 10
       time variable:  year, 1935 to 1954
```

We can obtain OLS parameter estimates for a linear model of invest on mvalue and kstock while allowing the standard errors (and variance–covariance matrix of the estimates) to be consistent when the disturbances from each observation are not independent. We want specifically for the standard errors to be robust to each company having a different variance of the disturbances and to each company's observations being correlated with those of the other companies through time.

This model is fitted in Stata by typing

```
. xtpcse invest mvalue kstock
Linear regression, correlated panels corrected standard errors (PCSEs)

Group variable:    company                 Number of obs      =        200
Time variable:     year                    Number of groups   =         10
Panels:            correlated (balanced)   Obs per group: min =         20
Autocorrelation:   no autocorrelation                     avg =         20
                                                          max =         20
Estimated covariances      =         55    R-squared          =     0.8124
Estimated autocorrelations =          0    Wald chi2(2)       =     637.41
Estimated coefficients     =          3    Prob > chi2        =     0.0000
```

	Coef.	Panel-corrected Std. Err.	z	P>|z|	[95% Conf. Interval]	
mvalue	.1155622	.0072124	16.02	0.000	.101426	.1296983
kstock	.2306785	.0278862	8.27	0.000	.1760225	.2853345
_cons	-42.71437	6.780965	-6.30	0.000	-56.00482	-29.42392

◁

▷ Example

xtgls will produce more efficient FGLS estimates of the models' parameters, but with the disadvantage that the standard error estimates are conditional on the estimated disturbance covariance. Beck and Katz (1995) argue that the improvement in power using FGLS with such data is small and that the standard error estimates from FGLS are unacceptably optimistic (anti-conservative).

The FGLS model is fitted by typing

```
. xtgls invest mvalue kstock, panels(correlated)
Cross-sectional time-series FGLS regression

Coefficients:  generalized least squares
Panels:        heteroskedastic with cross-sectional correlation
Correlation:   no autocorrelation

Estimated covariances      =         55    Number of obs      =        200
Estimated autocorrelations =          0    Number of groups   =         10
Estimated coefficients     =          3    Time periods       =         20
                                           Wald chi2(2)       =    3738.07
Log likelihood             = -879.4274     Prob > chi2        =     0.0000
```

invest	Coef.	Std. Err.	z	P>|z|	[95% Conf. Interval]	
mvalue	.1127515	.0022364	50.42	0.000	.1083683	.1171347
kstock	.2231176	.0057363	38.90	0.000	.2118746	.2343605
_cons	-39.84382	1.717563	-23.20	0.000	-43.21018	-36.47746

The coefficients between the two models are very close; the constants differ substantially, but we are generally not interested in the constant. As Beck and Katz observed, the standard errors for the FGLS model are 50 to 100% smaller than those for the OLS model with PCSE.

If we were also concerned about autocorrelation of the disturbances, we could specify `correlation(ar1)` to obtain a model with a common AR(1) parameter along with our other concerns about the disturbances.

```
. xtpcse invest mvalue kstock, correlation(ar1)
(note: estimates of rho outside [-1,1] bounded to be in the range [-1,1])

Prais-Winsten regression, correlated panels corrected standard errors (PCSEs)

Group variable:    company                 Number of obs      =        200
Time variable:     year                    Number of groups   =         10
Panels:            correlated (balanced)   Obs per group: min =         20
Autocorrelation:   common AR(1)                           avg =         20
                                                          max =         20
Estimated covariances      =        55     R-squared          =     0.5468
Estimated autocorrelations =         1     Wald chi2(2)       =      93.71
Estimated coefficients     =         3     Prob > chi2        =     0.0000
```

	Coef.	Panel-corrected Std. Err.	z	P>|z|	[95% Conf. Interval]	
mvalue	.0950157	.0129934	7.31	0.000	.0695492	.1204822
kstock	.306005	.0603718	5.07	0.000	.1876784	.4243317
_cons	-39.12569	30.50355	-1.28	0.200	-98.91154	20.66016
rho	.9059774					

The estimate of the autocorrelation parameter is high, .926, and the standard errors are larger than for the model without autocorrelation, which is to be expected if there is autocorrelation.

◁

▷ Example

Let's estimate panel-specific autocorrelation parameters and change the method of estimating the autocorrelation parameter to the one typically used to estimate autocorrelation in time-series analysis.

```
. xtpcse invest mvalue kstock, correlation(psar1) rhotype(tscorr)

Prais-Winsten regression, correlated panels corrected standard errors (PCSEs)

Group variable:    company                 Number of obs      =        200
Time variable:     year                    Number of groups   =         10
Panels:            correlated (balanced)   Obs per group: min =         20
Autocorrelation:   panel-specific AR(1)                   avg =         20
                                                          max =         20
Estimated covariances      =        55     R-squared          =     0.8670
Estimated autocorrelations =        10     Wald chi2(2)       =     444.53
Estimated coefficients     =         3     Prob > chi2        =     0.0000
```

	Coef.	Panel-corrected Std. Err.	z	P>|z|	[95% Conf. Interval]	
mvalue	.1052613	.0086018	12.24	0.000	.0884021	.1221205
kstock	.3386743	.0367568	9.21	0.000	.2666322	.4107163
_cons	-58.18714	12.63687	-4.60	0.000	-82.95496	-33.41933
rhos =	.5135627	.87017	.9023497	.63368	.8571502 ...	.8752707

Beck and Katz (1995, 121) make a case against estimating panel-specific AR parameters as opposed to a single AR parameter for all panels.

◁

▷ Example

We can also diverge from PCSEs to estimate standard errors that are panel-corrected, but only for panel-level heteroskedasticity; that is, each company has a different variance of the disturbances. Allowing also for autocorrelation, we would type

```
. xtpcse invest mvalue kstock, correlation(ar1) hetonly
(note: estimates of rho outside [-1,1] bounded to be in the range [-1,1])

Prais-Winsten regression, heteroskedastic panels corrected standard errors

Group variable:     company                 Number of obs      =        200
Time variable:      year                    Number of groups   =         10
Panels:             heteroskedastic (balanced)  Obs per group: min =       20
Autocorrelation:    common AR(1)                           avg =       20
                                                           max =       20
Estimated covariances      =        10     R-squared          =     0.5468
Estimated autocorrelations =         1     Wald chi2(2)       =      91.72
Estimated coefficients     =         3     Prob > chi2        =     0.0000
```

	Coef.	Het-corrected Std. Err.	z	P>|z|	[95% Conf. Interval]
mvalue	.0950157	.0130872	7.26	0.000	.0693653 .1206661
kstock	.306005	.061432	4.98	0.000	.1856006 .4264095
_cons	-39.12569	26.16935	-1.50	0.135	-90.41666 12.16529
rho	.9059774				

With this specification, we are *NOT* obtaining what are referred to in the literature as PCSEs. These standard errors are in the same spirit as PCSEs, but are from the asymptotic covariance estimates of OLS without allowing for contemporaneous correlation.

◁

(*Continued on next page*)

Saved Results

xtpcse saves in e():

Scalars

e(N)	number of observations	e(rmse)	root mean square error
e(N_g)	number of groups	e(g_max)	largest group size
e(n_cf)	number of estimated coefficients	e(g_min)	smallest group size
e(n_cv)	number of estimated covariances	e(g_avg)	average group size
e(n_cr)	number of estimated correlations	e(rc)	return code
e(mss)	model sum of squares	e(chi2)	χ^2
e(df)	degrees of freedom	e(p)	significance
e(df_m)	model degrees of freedom	e(N_gaps)	number of gaps
e(rss)	residual sum of squares	e(n_sigma)	obs used to estimate elements
e(r2)	R-squared		of Sigma

Macros

e(cmd)	xtpcse	e(vcetype)	covariance estimation method
e(depvar)	name of dependent variable	e(chi2type)	Wald; type of model χ^2 test
e(title)	title in estimation output	e(rho)	ρ
e(panels)	contemporaneous covariance structure	e(cons)	noconstant or ""
e(corr)	correlation structure	e(missmeth)	casewise or pairwise
e(rhotype)	type of estimated correlation	e(balance)	balanced or unbalanced
e(ivar)	variable denoting groups	e(predict)	program used to implement
e(tvar)	variable denoting time		predict

Matrices

e(b)	coefficient vector	e(Sigma)	$\widehat{\Sigma}$ matrix
e(V)	variance–covariance matrix	e(rhomat)	vector of autocorrelation
	of the estimators		parameter estimates

Functions

e(sample)	marks estimation sample

Methods and Formulas

xtpcse is implemented as an ado-file.

If no autocorrelation is specified, the parameters β are estimated by OLS; see [R] **regress**. If autocorrelation is specified, the parameters β are estimated by Prais–Winsten; see [TS] **prais**.

When autocorrelation with panel-specific coefficients of correlation is specified (option correlation(psar1)), each panel-level ρ_i is computed from the residuals of an OLS regression across all panels; see [TS] **prais**. When autocorrelation with a common coefficient of correlation is specified (option correlation(ar1)), the common correlation coefficient is computed as

$$\rho = \frac{\rho_1 + \rho_2 + \cdots + \rho_m}{m}$$

where ρ_i is the estimated autocorrelation coefficient for panel i and m is the number of panels.

The covariance of the OLS or Prais–Winsten coefficients is

$$\text{Var}(\boldsymbol{\beta}) = (\mathbf{X}'\mathbf{X})^{-1}\mathbf{X}'\boldsymbol{\Omega}\mathbf{X}(\mathbf{X}'\mathbf{X})^{-1}$$

where $\boldsymbol{\Omega}$ is the full covariance matrix of the disturbances.

When the panels are balanced, we can write $\boldsymbol{\Omega}$ as

$$\boldsymbol{\Omega} = \boldsymbol{\Sigma}_{m \times m} \otimes \mathbf{I}_{T_i \times T_i}$$

where $\boldsymbol{\Sigma}$ is the m by m panel-by-panel covariance matrix of the disturbances; see *Remarks*.

xtpcse estimates the elements of $\boldsymbol{\Sigma}$ as

$$\widehat{\boldsymbol{\Sigma}}_{ij} = \frac{\epsilon_i{}'\epsilon_j}{T_{ij}}$$

where ϵ_i and ϵ_j are the residuals for panels i and j, respectively, that can be matched by time period, and where T_{ij} is the number of residuals between the panels i and j that can be matched by time period.

When the panels are balanced (each panel has the same number of observations and all time periods are common to all panels), then $T_{ij} = T$, where T is the number of observations per panel.

When panels are unbalanced, xtpcse by default uses casewise selection, where only those residuals from time periods that are common to all panels are used to compute $\widehat{S}_{ij}$. In this case, $T_{ij} = T^*$, where T^* is the number of time periods common to all panels. When pairwise is specified, each $\widehat{S}_{ij}$ is computed using all observations that can be matched by time period between the panels i and j.

Acknowledgment

We would like to thank the following people for helpful comments: Nathaniel Beck, Department of Political Science, University of California at San Diego; Jonathan Katz, Division of the Humanities and Social Science, California Institute of Technology; and Robert John Franzese, Jr., Center for Political Studies, Institute for Social Research, University of Michigan.

References

Beck, N. and J. N. Katz. 1995. What to do (and not to do) with time-series cross-section data. *American Political Science Review* 89: 634–647.

Davidson, R. and J. G. MacKinnon. 1993. *Estimation and Inference in Econometrics*. New York: Oxford University Press.

Greene, W. H. 2003. *Econometric Analysis*. 5th ed. Upper Saddle River, NJ: Prentice–Hall.

Grunfeld, Y. and Z. Griliches. 1960. Is aggregation necessarily bad? *Review of Economics and Statistics* 42: 1–13.

Judge, G. G., W. E. Griffiths, R. C. Hill, H. Lütkepohl, and T.-C. Lee. 1985. *The Theory and Practice of Econometrics*. 2d ed. New York: John Wiley & Sons.

Kmenta, J. 1997. *Elements of Econometrics*. 2d ed. Ann Arbor: University of Michigan Press.

Also See

Complementary:	[XT] **xtdata**, [XT] **xtdes**, [XT] **xtsum**, [XT] **xttab**, [R] **adjust**, [R] **lincom**, [R] **mfx**, [R] **nlcom**, [R] **predict**, [R] **predictnl**, [R] **test**, [R] **testnl**, [R] **vce**, [TS] **tsset**
Related:	[XT] **xtgee**, [XT] **xtgls**, [XT] **xtreg**, [XT] **xtregar**, [R] **regress**, [SVY] **svy estimators**, [TS] **newey**, [TS] **prais**
Background:	[U] **16.5 Accessing coefficients and standard errors**, [U] **23 Estimation and post-estimation commands**, [XT] **xt**

Title

> **xtpoisson** — Fixed-effects, random-effects, and population-averaged Poisson models

Syntax

Random-effects model

> xtpoisson *depvar* [*varlist*] [*weight*] [if *exp*] [in *range*] [, re i(*varname*) irr
>
> quad(*#*) constraints(*numlist*) noskip noconstant level(*#*) normal
>
> exposure(*varname*) offset(*varname*) nolog *maximize_options*]

Conditional fixed-effects model

> xtpoisson *depvar* [*varlist*] [*weight*] [if *exp*] [in *range*] , fe [i(*varname*) irr
>
> constraints(*numlist*) noskip level(*#*) exposure(*varname*) offset(*varname*)
>
> nolog *maximize_options*]

Population-averaged model

> xtpoisson *depvar* [*varlist*] [*weight*] [if *exp*] [in *range*] , pa [i(*varname*) irr
>
> noconstant level(*#*) robust exposure(*varname*) offset(*varname*)
>
> nolog *xtgee_options* *maximize_options*]

by ... : may be used with xtpoisson; see [R] **by**.

iweights, fweights, and pweights are allowed for the population-averaged model and iweights are allowed in the random-effects and fixed-effects models; see [U] **14.1.6 weight**. Note that weights must be constant within panels.

xtpoisson shares the features of all estimation commands; see [U] **23 Estimation and post-estimation commands**.

Syntax for predict

Random-effects and fixed-effects models

> predict [*type*] *newvarname* [if *exp*] [in *range*] [, [xb | stdp | nu0 | iru0]
>
> nooffset]

Population-averaged model

> predict [*type*] *newvarname* [if *exp*] [in *range*] [, [mu | rate | xb | stdp]
>
> nooffset]

These statistics are available both in and out of sample; type predict ... if e(sample) ... if wanted only for the estimation sample.

Description

    xtpoisson fits random-effects, conditional fixed-effects, and population-averaged Poisson models. Whenever we refer to a fixed-effects model, we mean the conditional fixed-effects model.

    Note: xtpoisson, re normal is slow since it is calculated by quadrature; see *Methods and Formulas*. Computation time is roughly proportional to the number of points used for the quadrature. The default is quad(12). Simulations indicate that increasing it does not appreciably change the estimates for the coefficients or their standard errors. See [XT] **quadchk**.

    By default, the population-averaged model is an equal-correlation model; xtpoisson assumes corr(exchangeable). See [XT] **xtgee** for information on how to fit other population-averaged models.

Options

re, the default, requests the random-effects estimator.

fe requests the fixed-effects estimator.

pa requests the population-averaged estimator.

i(*varname*) specifies the variable name that contains the unit to which the observation belongs. You can specify the i() option the first time you estimate, or you can use the iis command to set i() beforehand. Note that it is not necessary to specify i() if the data have been previously tsset, or if iis has been previously specified—in these cases, the group variable is taken from the previous setting. See [XT] **xt**.

irr reports exponentiated coefficients e^b rather than coefficients b. For the Poisson model, exponentiated coefficients have the interpretation of incidence-rate ratios.

quad(*#*) specifies the number of points to use in the quadrature approximation of the integral. This option is relevant only if you are fitting a random-effects model; if you specify quad(*#*) with pa, the quad(*#*) is ignored.

    The default is quad(12). The number specified must be an integer between 4 and 30, and must be no greater than the number of observations.

constraints(*numlist*) specifies by number the linear constraints to be applied during estimation. The default is to perform unconstrained estimation. Constraints are specified using the constraint command; see [R] **constraint**. constraints(*numlist*) may not be specified with pa.

noskip specifies that a full maximum-likelihood model with only a constant for the regression equation be fitted. This model is not displayed, but is used as the base model to compute a likelihood-ratio test for the model test statistic displayed in the estimation header. By default, the overall model test statistic is an asymptotically equivalent Wald test of all the parameters in the regression equation being zero (except the constant). For many models, this option can substantially increase estimation time.

noconstant suppresses the constant term (intercept) in the model.

level(*#*) specifies the confidence level, in percent, for confidence intervals. The default is level(95) or as set by set level; see [U] **23.6 Specifying the width of confidence intervals**.

normal specifies that the random effects follow a normal distribution instead of a gamma distribution.

exposure(*varname*) and offset(*varname*) are different ways of specifying the same thing. exposure() specifies a variable that reflects the amount of exposure over which the *depvar* events were observed for each observation; ln(*varname*) with its coefficient constrained to be 1 is entered into the regression equation. offset() specifies a variable that is to be entered directly into the regression equation with its coefficient constrained to be 1; thus, exposure is assumed to be $e^{varname}$.

robust specifies that the Huber/White/sandwich estimator of variance is to be used in place of the IRLS variance estimator; see [XT] **xtgee**. This alternative produces valid standard errors even if the correlations within group are not as hypothesized by the specified correlation structure. It does, however, require that the model correctly specifies the mean. As such, the resulting standard errors are labeled "semi-robust" instead of "robust". Note that although there is no cluster() option, results are as if there were a cluster() option and you specified clustering on i().

nolog suppresses the iteration log.

xtgee_options specifies any other options allowed by xtgee for family(poisson) link(log); see [XT] **xtgee**.

maximize_options control the maximization process; see [R] **maximize**. Use the trace option to view parameter convergence. Use the ltol(#) option to relax the convergence criterion; the default is 1e−6 during specification searches.

Options for predict

xb calculates the linear prediction. This is the default for the random-effects and fixed-effects models.

mu and rate both calculate the predicted probability of *depvar*. mu takes into account the offset(), and rate ignores those adjustments. mu and rate are equivalent if you did not specify offset(). mu is the default for the population-averaged model.

stdp calculates the standard error of the linear prediction.

nu0 calculates the predicted number of events assuming a zero random/fixed effect.

iru0 calculates the predicted incidence rate assuming a zero random/fixed effect.

nooffset is relevant only if you specified offset(*varname*) for xtpoisson. It modifies the calculations made by predict so that they ignore the offset variable; the linear prediction is treated as $\mathbf{x}_{it}\boldsymbol{\beta}$ rather than $\mathbf{x}_{it}\boldsymbol{\beta} + \text{offset}_{it}$.

Remarks

xtpoisson is a convenience command if you want the population-averaged model. Typing

    . xtpoisson ..., ... pa exposure(time)

is equivalent to typing

    . xtgee ..., ... family(poisson) link(log) corr(exchangeable) exposure(time)

Thus, also see [XT] **xtgee** for information about xtpoisson.

By default, or when re is specified, xtpoisson fits via maximum-likelihood the random-effects model

$$\Pr(Y_{it} = y_{it}|\mathbf{x}_{it}) = F(y_{it}, \mathbf{x}_{it}\boldsymbol{\beta} + \nu_i)$$

for $i = 1, \ldots, n$ panels, $t = 1, \ldots, n_i$, and $F(x, z) = \Pr(X = x)$ where X is Poisson distributed with mean $\exp(z)$. In the standard random-effects model, ν_i is assumed to be iid such that $\exp(\nu_i)$ is gamma with mean one and variance α, which is estimated from the data. If normal is specified, then ν_i is assumed to be iid $N(0, \sigma_\nu^2)$.

▷ Example

You have data on the number of ship accidents for 5 different types of ships (McCullagh and Nelder 1989, 205). You wish to analyze whether the "incident" rate is affected by the period in which the ship was constructed and operated. Your measure of exposure is months of service for the ship, and in this model, we assume that the exponentiated random effects are distributed as gamma with mean one and variance α.

```
. use http://www.stata-press.com/data/r8/ships

. xtpoisson accident op_75_79 co_65_69 co_70_74 co_75_79, i(ship) ex(service) irr

Fitting comparison Poisson model:

Iteration 0:   log likelihood = -147.37993
Iteration 1:   log likelihood = -80.372714
Iteration 2:   log likelihood = -80.116093
Iteration 3:   log likelihood = -80.115916
Iteration 4:   log likelihood = -80.115916

Fitting full model:

Iteration 0:   log likelihood = -79.653186
Iteration 1:   log likelihood = -76.990836  (not concave)
Iteration 2:   log likelihood = -74.824942
Iteration 3:   log likelihood = -74.811243
Iteration 4:   log likelihood = -74.811217
Iteration 5:   log likelihood = -74.811217
```

```
Random-effects Poisson regression        Number of obs      =         34
Group variable (i) : ship                Number of groups   =          5
Random effects u_i ~ Gamma               Obs per group: min =          6
                                                        avg =        6.8
                                                        max =          7
                                         Wald chi2(4)       =      50.90
Log likelihood  = -74.811217             Prob > chi2        =     0.0000
```

accident	IRR	Std. Err.	z	P>|z|	[95% Conf. Interval]	
op_75_79	1.466305	.1734005	3.24	0.001	1.162957	1.848777
co_65_69	2.032543	.304083	4.74	0.000	1.515982	2.72512
co_70_74	2.356853	.3999259	5.05	0.000	1.690033	3.286774
co_75_79	1.641913	.3811398	2.14	0.033	1.04174	2.58786
service	(exposure)					
/lnalpha	-2.368406	.8474597			-4.029397	-.7074155
alpha	.0936298	.0793475			.0177851	.4929165

```
Likelihood-ratio test of alpha=0: chibar2(01) =    10.61 Prob>=chibar2 = 0.001
```

The output also includes a likelihood-ratio test of $\alpha = 0$, which compares the panel estimator with the pooled (Poisson) estimator.

You find that the incidence rate for accidents is significantly different for the periods of construction and operation of the ships, and that the random-effects model is significantly different from the pooled model.

We may alternatively fit a fixed-effects specification instead of a random-effects specification:

```
. xtpoisson accident op_75_79 co_65_69 co_70_74 co_75_79, i(ship) ex(service) irr fe

Iteration 0:   log likelihood = -80.738973
Iteration 1:   log likelihood = -54.857546
Iteration 2:   log likelihood = -54.641897
Iteration 3:   log likelihood = -54.641859
Iteration 4:   log likelihood = -54.641859

Conditional fixed-effects Poisson regression     Number of obs     =         34
Group variable (i): ship                         Number of groups  =          5

                                                 Obs per group: min =          6
                                                                avg =        6.8
                                                                max =          7

                                                 Wald chi2(4)      =      48.44
Log likelihood  = -54.641859                     Prob > chi2       =     0.0000
```

accident	IRR	Std. Err.	z	P>|z|	[95% Conf. Interval]	
op_75_79	1.468831	.1737218	3.25	0.001	1.164926	1.852019
co_65_69	2.008003	.3004803	4.66	0.000	1.497577	2.692398
co_70_74	2.26693	.384865	4.82	0.000	1.625274	3.161912
co_75_79	1.573695	.3669393	1.94	0.052	.9964273	2.485397
service	(exposure)					

Both of these models estimate the same thing. The difference will be in efficiency, depending on whether the assumptions of the random-effects model are true.

Note that we could have assumed that the random effects followed a normal distribution, $N(0, \sigma_\nu^2)$, instead of a "log-gamma" distribution, and obtained

```
. xtpoisson accident op_75_79 co_65_69 co_70_74 co_75_79, i(ship) ex(service) irr
> normal nolog

Random-effects Poisson regression                Number of obs     =         34
Group variable (i): ship                         Number of groups  =          5

Random effects u_i ~ Gaussian                    Obs per group: min =          6
                                                                avg =        6.8
                                                                max =          7

                                                 LR chi2(4)        =      55.93
Log likelihood  = -74.225924                     Prob > chi2       =     0.0000
```

accident	IRR	Std. Err.	z	P>|z|	[95% Conf. Interval]	
op_75_79	1.470182	.1737941	3.26	0.001	1.166134	1.853505
co_65_69	2.025867	.3030042	4.72	0.000	1.511119	2.715958
co_70_74	2.336483	.3960786	5.01	0.000	1.675975	3.257299
co_75_79	1.640625	.3777041	2.15	0.032	1.044831	2.576159
service	(exposure)					
/lnsig2u	-1.42662	.5613872	-2.54	0.011	-2.526919	-.3263217
sigma_u	.4900195	.1375453			.2826744	.8494545

Likelihood-ratio test of sigma_u=0: chibar2(01) =   11.78 Pr>=chibar2 = 0.000

The output includes the additional panel-level variance component. This is parameterized as the log of the variance $\ln(\sigma_\nu^2)$ (labeled lnsig2u in the output). The standard deviation σ_ν is also included in the output labeled sigma_u.

When `sigma_u` is zero, the panel-level variance component is unimportant and the panel estimator is not different from the pooled estimator. A likelihood-ratio test of this is included at the bottom of the output. This test formally compares the pooled estimator (poisson) with the panel estimator. In this case, `sigma_u` is not zero, so a panel estimator is indicated.

◁

▷ Example

We rerun our previous example, but this time we fit a robust equal-correlation population-averaged model:

```
. xtpoisson accident op_75_79 co_65_69 co_70_74 co_75_79, i(ship) ex(service)
> pa robust eform

Iteration 1: tolerance = .04083192
Iteration 2: tolerance = .00270188
Iteration 3: tolerance = .00030663
Iteration 4: tolerance = .00003466
Iteration 5: tolerance = 3.891e-06
Iteration 6: tolerance = 4.359e-07

GEE population-averaged model          Number of obs     =         34
Group variable:                ship    Number of groups  =          5
Link:                           log    Obs per group: min =         6
Family:                     Poisson                   avg =        6.8
Correlation:            exchangeable                  max =          7
                                       Wald chi2(3)      =     181.55
Scale parameter:                  1    Prob > chi2       =     0.0000

                 (standard errors adjusted for clustering on ship)
```

accident	IRR	Semi-robust Std. Err.	z	P>|z|	[95% Conf. Interval]	
op_75_79	1.483299	.1197901	4.88	0.000	1.266153	1.737685
co_65_69	2.038477	.1809524	8.02	0.000	1.712955	2.425859
co_70_74	2.643467	.4093947	6.28	0.000	1.951407	3.580962
co_75_79	1.876656	.33075	3.57	0.000	1.328511	2.650966
service	(exposure)					

We may compare this with a pooled estimator with clustered robust-variance estimates:

(Continued on next page)

```
. poisson accident op_75_79 co_65_69 co_70_74 co_75_79, ex(service) robust
> cluster(ship) irr

Iteration 0:   log likelihood = -147.37993
Iteration 1:   log likelihood = -80.372714
Iteration 2:   log likelihood = -80.116093
Iteration 3:   log likelihood = -80.115916
Iteration 4:   log likelihood = -80.115916
```

Poisson regression   Number of obs =   34
                                                    Wald chi2(3)   =   .
Log pseudo-likelihood = -80.115916   Prob > chi2   =   .

                          (standard errors adjusted for clustering on ship)

accident	IRR	Robust Std. Err.	z	P>|z|	[95% Conf. Interval]	
op_75_79	1.47324	.1287036	4.44	0.000	1.2414	1.748377
co_65_69	2.125914	.2850531	5.62	0.000	1.634603	2.764897
co_70_74	2.860138	.6213563	4.84	0.000	1.868384	4.378325
co_75_79	2.021926	.4265285	3.34	0.001	1.337221	3.057227
service	(exposure)					

◁

Saved Results

xtpoisson, re saves in e():

Scalars

e(N)	number of observations	e(g_min)	smallest group size
e(k)	number of variables	e(g_avg)	average group size
e(k_eq)	number of equations	e(rc)	return code
e(k_dv)	number of dependent variables	e(chi2)	χ^2
e(N_g)	number of groups	e(chi2_c)	χ^2 for comparison test
e(df_m)	model degrees of freedom	e(p)	significance
e(ll)	log likelihood	e(ic)	number of iterations
e(ll_0)	log likelihood, constant-only model	e(n_quad)	number of quadrature points
e(ll_c)	log likelihood, comparison model	e(alpha)	α
e(g_max)	largest group size		

Macros

e(cmd)	xtpoisson	e(opt)	type of optimization
e(depvar)	name of dependent variable	e(chi2type)	Wald or LR; type of model χ^2 test
e(title)	title in estimation output	e(chi2_ct)	Wald or LR; type of model χ^2 test
e(ivar)	variable denoting groups		corresponding to e(chi2_c)
e(wtype)	weight type	e(distrib)	Gamma; the distribution of the
e(wexp)	weight expression		random effect
e(user)	name of likelihood-evaluator program	e(crittype)	optimization criterion
e(offset)	offset	e(predict)	program used to implement predict

Matrices

e(b)	coefficient vector	e(V)	variance–covariance matrix of the
e(ilog)	iteration log		estimators

Functions

e(sample)	marks estimation sample

xtpoisson, re normal saves in e():

Scalars

e(N)	number of observations	e(g_min)	smallest group size
e(N_g)	number of groups	e(g_avg)	average group size
e(df_m)	model degrees of freedom	e(chi2)	χ^2
e(ll)	log likelihood	e(chi2_c)	χ^2 for comparison test
e(ll_0)	log likelihood, constant-only model	e(sigma_u)	panel-level standard deviation
e(ll_c)	log likelihood, comparison model	e(N_cd)	number of completely determined obs.
e(g_max)	largest group size	e(n_quad)	number of quadrature points

Macros

e(cmd)	xtpoisson	e(chi2type)	Wald or LR; type of model χ^2 test
e(depvar)	name of dependent variable	e(chi2_ct)	Wald or LR; type of model χ^2 test
e(title)	title in estimation output		corresponding to e(chi2_c)
e(ivar)	variable denoting groups	e(distrib)	Gaussian; the distribution of the
e(wtype)	weight type		random effect
e(wexp)	weight expression	e(crittype)	optimization criterion
e(offset1)	offset	e(predict)	program used to implement predict

Matrices

e(b)	coefficient vector	e(V)	variance–covariance matrix of the estimators

Functions

e(sample)	marks estimation sample

xtpoisson, fe saves in e():

Scalars

e(N)	number of observations	e(g_max)	largest group size
e(k)	number of variables	e(g_min)	smallest group size
e(k_eq)	number of equations	e(g_avg)	average group size
e(k_dv)	number of dependent variables	e(rc)	return code
e(N_g)	number of groups	e(chi2)	χ^2
e(df_m)	model degrees of freedom	e(p)	significance
e(ll)	log likelihood	e(ic)	number of iterations
e(ll_0)	log likelihood, constant-only model	e(n_quad)	number of quadrature points

Macros

e(cmd)	xtpoisson	e(user)	name of likelihood-evaluator program
e(depvar)	name of dependent variable	e(opt)	type of optimization
e(title)	title in estimation output	e(chi2type)	LR; type of model χ^2 test
e(ivar)	variable denoting groups	e(offset)	offset
e(wtype)	weight type	e(crittype)	optimization criterion
e(wexp)	weight expression	e(predict)	program used to implement predict

Matrices

e(b)	coefficient vector	e(V)	variance–covariance matrix of the estimators

Functions

e(sample)	marks estimation sample

`xtpoisson, pa` saves in `e()`:

Scalars

`e(N)`	number of observations	`e(deviance)`	deviance
`e(N_g)`	number of groups	`e(chi2_dev)`	χ^2 test of deviance
`e(df_m)`	model degrees of freedom	`e(dispers)`	deviance dispersion
`e(g_max)`	largest group size	`e(chi2_dis)`	χ^2 test of deviance dispersion
`e(g_min)`	smallest group size	`e(tol)`	target tolerance
`e(g_avg)`	average group size	`e(dif)`	achieved tolerance
`e(chi2)`	χ^2	`e(phi)`	scale parameter
`e(df_pear)`	degrees of freedom for Pearson χ^2		

Macros

`e(cmd)`	xtgee	`e(scale)`	x2, dev, phi, or #; scale parameter
`e(cmd2)`	xtpoisson	`e(ivar)`	variable denoting groups
`e(depvar)`	name of dependent variable	`e(vcetype)`	covariance estimation method
`e(family)`	Poisson	`e(chi2type)`	Wald; type of model χ^2 test
`e(link)`	log; link function	`e(offset)`	offset
`e(corr)`	correlation structure	`e(predict)`	program used to implement predict
`e(crittype)`	optimization criterion		

Matrices

`e(b)`	coefficient vector	`e(V)`	variance–covariance matrix of the estimators
`e(R)`	estimated working correlation matrix		

Functions

`e(sample)`	marks estimation sample

Methods and Formulas

`xtpoisson` is implemented as an ado-file.

`xtpoisson, pa` reports the population-averaged results obtained by using `xtgee, family(poisson) link(log)` to obtain estimates. See [XT] **xtgee** for details on the methods and formulas.

While Hausman, Hall, and Griliches (1984) is the seminal article on the random-effects and fixed-effects models, Cameron and Trivedi (1998) has a good textbook treatment.

For a random-effects specification, we know that

$$\Pr(y_{i1}, \ldots, y_{in_i} | \alpha_i, \mathbf{x}_{i1}, \ldots, \mathbf{x}_{in_i}) = \left(\prod_{t=1}^{n_i} \frac{\lambda_{it}^{y_{it}}}{y_{it}!} \right) \exp \left\{ -\exp(\alpha_i) \sum_{t=1}^{n_i} \lambda_{it} \right\} \exp \left(\alpha_i \sum_{t=1}^{n_i} y_{it} \right)$$

where $\lambda_{it} = \exp(\mathbf{x}_{it}\boldsymbol{\beta})$. We may rewrite the above as (defining $\epsilon_i = \exp(\alpha_i)$)

(Continued on next page)

$$\Pr(y_{i1}, \ldots, y_{in_i} | \epsilon_i, \mathbf{x}_{i1}, \ldots, \mathbf{x}_{in_i}) = \left\{ \prod_{t=1}^{n_i} \frac{(\lambda_{it}\epsilon_i)^{y_{it}}}{y_{it}!} \right\} \exp\left(-\sum_{t=1}^{n_i} \lambda_{it}\epsilon_i \right)$$

$$= \left(\prod_{t=1}^{n_i} \frac{\lambda_{it}^{y_{it}}}{y_{it}!} \right) \exp\left(-\epsilon_i \sum_{t=1}^{n_i} \lambda_{it} \right) \epsilon_i^{\sum_{t=1}^{n_i} y_{it}}$$

We now assume that ϵ_i follows a gamma distribution with mean one and variance θ so that unconditional on ϵ_i

$$\Pr(y_{i1}, \ldots, y_{in_i} | \mathbf{X}_i) = \frac{\theta^\theta}{\Gamma(\theta)} \left(\prod_{t=1}^{n_i} \frac{\lambda_{it}^{y_{it}}}{y_{it}!} \right) \int_0^\infty \exp\left(-\epsilon_i \sum_{t=1}^{n_i} \lambda_{it} \right) \epsilon_i^{\sum_{t=1}^{n_i} y_{it}} \epsilon_i^{\theta-1} \exp(-\theta\epsilon_i) d\epsilon_i$$

$$= \frac{\theta^\theta}{\Gamma(\theta)} \left(\prod_{t=1}^{n_i} \frac{\lambda_{it}^{y_{it}}}{y_{it}!} \right) \int_0^\infty \exp\left\{ -\epsilon_i \left(\theta + \sum_{t=1}^{n_i} \lambda_{it} \right) \right\} \epsilon_i^{\theta + \sum_{t=1}^{n_i} y_{it} - 1} d\epsilon_i$$

$$= \left(\prod_{t=1}^{n_i} \frac{\lambda_{it}^{y_{it}}}{y_{it}!} \right) \frac{\Gamma\left(\theta + \sum_{t=1}^{n_i} y_{it} \right)}{\Gamma(\theta)} \left(\frac{\theta}{\theta + \sum_{t=1}^{n_i} \lambda_{it}} \right)^\theta \left(\frac{1}{\theta + \sum_{t=1}^{n_i} \lambda_{it}} \right)^{\sum_{t=1}^{n_i} y_{it}}$$

for $\mathbf{X}_i = (\mathbf{x}_{i1}, \ldots, \mathbf{x}_{in_i})$.

The log likelihood (assuming Gamma heterogeneity) is then derived using

$$u_i = \frac{\theta}{\theta + \sum_{t=1}^{n_i} \lambda_{it}} \qquad \lambda_{it} = \exp(\mathbf{x}_{it}\boldsymbol{\beta})$$

$$\Pr(Y_{i1} = y_{i1}, \ldots, Y_{in_i} = y_{in_i} | \mathbf{X}_i) = \frac{\prod_{t=1}^{n_i} \lambda_{it}^{y_{it}} \Gamma\left(\theta + \sum_{t=1}^{n_i} y_{it} \right)}{\prod_{t=1}^{n_i} y_{it}! \Gamma(\theta) \left(\sum_{t=1}^{n_i} \lambda_{it} \right)^{\sum_{t=1}^{n_i} y_{it}}} u_i^\theta (1 - u_i)^{\sum_{t=1}^{n_i} y_{it}}$$

such that the log likelihood may be written as

$$L = \sum_{i=1}^n w_i \left\{ \log\Gamma\left(\theta + \sum_{t=1}^{n_i} y_{it} \right) - \sum_{t=1}^{n_i} \log\Gamma\left(1 + y_{it} \right) - \log\Gamma(\theta) + \theta\log u_i \right.$$

$$\left. + \log(1 - u_i) \sum_{t=1}^{n_i} y_{it} + \sum_{t=1}^{n_i} y_{it}(\mathbf{x}_{it}\boldsymbol{\beta}) - \left(\sum_{t=1}^{n_i} y_{it} \right) \log\left(\sum_{t=1}^{n_i} \lambda_{it} \right) \right\}$$

where w_i is the user-specified weight for panel i; if no weights are specified, $w_i = 1$.

(*Continued on next page*)

Alternatively, if we assume a normal distribution, $N(0, \sigma_\nu^2)$, for the random effects ν_i, we have that

$$\Pr(y_{i1}, \ldots, y_{in_i} | \mathbf{X}_i) = \int_{-\infty}^{\infty} \frac{e^{-\nu_i^2/2\sigma_\nu^2}}{\sqrt{2\pi}\sigma_\nu} \left\{ \prod_{t=1}^{n_i} F(y_{it}, \mathbf{x}_{it}\boldsymbol{\beta} + \nu_i) \right\} d\nu_i$$

where

$$F(y, z) = \exp\left\{ -\exp(z) + yz - \log(y!) \right\}$$

and we can approximate the integral with M-point Gauss–Hermite quadrature

$$\int_{-\infty}^{\infty} e^{-x^2} g(x) dx \approx \sum_{m=1}^{M} w_m^* g(a_m^*)$$

where the w_m^* denote the quadrature weights and the a_m^* denote the quadrature abscissas. The log-likelihood L, where $\tau = \sigma_\nu^2/(\sigma_\nu^2 + 1)$, is then calculated using the quadrature

$$L = \sum_{i=1}^{n} w_i \log\left\{ \Pr(y_{i1}, \ldots, y_{in_i} | \mathbf{X}_i) \right\}$$

$$\approx \sum_{i=1}^{n} w_i \log\left[\frac{1}{\sqrt{\pi}} \sum_{m=1}^{M} w_m^* \prod_{t=1}^{n_i} F\left\{ y_{it}, \mathbf{x}_{it}\boldsymbol{\beta} + a_m^* \left(\frac{2\tau}{1-\tau} \right)^{1/2} \right\} \right]$$

The quadrature formula requires that the integrated function be well-approximated by a polynomial. As the number of time periods becomes large (as panel size gets large),

$$\prod_{t=1}^{n_i} F(y_{it}, \mathbf{x}_{it}\boldsymbol{\beta} + \nu_i)$$

is no longer well-approximated by a polynomial. As a general rule of thumb, you should use this quadrature approach only for small to moderate panel sizes (based on simulations, 50 is a reasonably safe upper bound). However, if the data really come from random-effects poisson and τ is not too large (less than, say, .3), then the panel size could be 500 and the quadrature approximation would still be fine. If the data are not random-effects poisson or τ is large (bigger than, say, .7), then the quadrature approximation may be poor for panel sizes larger than 10. The quadchk command should be used to investigate the applicability of the numeric technique used in this command.

For a fixed-effects specification, we know that

$$\Pr(Y_{it} = y_{it} | \mathbf{x}_{it}) = \exp\{ -\exp(\alpha_i + \mathbf{x}_{it}\boldsymbol{\beta}) \} \exp(\alpha_i + \mathbf{x}_{it}\boldsymbol{\beta})^{y_{it}} / y_{it}!$$

$$= \frac{1}{y_{it}!} \exp\{ -\exp(\alpha_i) \exp(\mathbf{x}_{it}\boldsymbol{\beta}) + \alpha_i y_{it} \} \exp(\mathbf{x}_{it}\boldsymbol{\beta})^{y_{it}}$$

$$\equiv F_{it}$$

Since we know that the observations are independent, we may write the joint probability for the observations within a panel as

$$\Pr(Y_{i1} = y_{i1}, \ldots, Y_{in_i} = y_{in_i} | \mathbf{X}_i) = \prod_{t=1}^{n_i} \frac{1}{y_{it}!} \exp\{-\exp(\alpha_i)\exp(\mathbf{x}_{it}\boldsymbol{\beta}) + \alpha_i y_{it}\} \exp(\mathbf{x}_{it}\boldsymbol{\beta})^{y_{it}}$$

$$= \left(\prod_{t=1}^{n_i} \frac{\exp(\mathbf{x}_{it}\boldsymbol{\beta})^{y_{it}}}{y_{it}!} \right) \exp\left\{ -\exp(\alpha_i) \sum_t \exp(\mathbf{x}_{it}\boldsymbol{\beta}) + \alpha_i \sum_t y_{it} \right\}$$

and we also know that the sum of n_i Poisson independent random variables, each with parameter λ_{it} for $t = 1, \ldots, n_i$, is distributed as Poisson with parameter $\sum_t \lambda_{it}$. Thus

$$\Pr\left(\sum_t Y_{it} = \sum_t y_{it} \Big| \mathbf{X}_i \right) =$$

$$\frac{1}{(\sum_t y_{it})!} \exp\left\{ -\exp(\alpha_i)\sum_t \exp(\mathbf{x}_{it}\boldsymbol{\beta}) + \alpha_i \sum_t y_{it} \right\} \left\{ \sum_t \exp(\mathbf{x}_{it}\boldsymbol{\beta}) \right\}^{\sum_t y_{it}}$$

So, the conditional likelihood is conditioned on the sum of the outcomes in the set (panel). The appropriate function is given by

$$\Pr\left(Y_{i1} = y_{i1}, \ldots, Y_{in_i} = y_{in_i} \Big| \mathbf{X}_i, \sum_t Y_{it} = \sum_t y_{it} \right) =$$

$$\left[\left(\prod_{t=1}^{n_i} \frac{\exp(\mathbf{x}_{it}\boldsymbol{\beta})^{y_{it}}}{y_{it}!} \right) \exp\left\{ -\exp(\alpha_i)\sum_t \exp(\mathbf{x}_{it}\boldsymbol{\beta}) + \alpha_i \sum_t y_{it} \right\} \right] \Big/$$

$$\left[\frac{1}{(\sum_t y_{it})!} \exp\left\{ -\exp(\alpha_i)\sum_t \exp(\mathbf{x}_{it}\boldsymbol{\beta}) + \alpha_i \sum_t y_{it} \right\} \left\{ \sum_t \exp(\mathbf{x}_{it}\boldsymbol{\beta}) \right\}^{\sum_t y_{it}} \right]$$

$$= \left(\sum_t y_{it} \right)! \prod_{t=1}^{n_i} \frac{\exp(\mathbf{x}_{it}\boldsymbol{\beta})^{y_{it}}}{y_{it}! \left\{ \sum_k \exp(\mathbf{x}_{ik}\boldsymbol{\beta}) \right\}^{y_{it}}}$$

which is free of α_i.

So, the conditional log-likelihood is given by

(Continued on next page)

$$L = \log \prod_{i=1}^{n} \left[\left(\sum_{t=1}^{n_i} y_{it} \right)! \prod_{t=1}^{n_i} \frac{\exp(\mathbf{x}_{it}\boldsymbol{\beta})^{y_{it}}}{y_{it}! \left\{ \sum_{\ell=1}^{n_\ell} \exp(\mathbf{x}_{i\ell}\boldsymbol{\beta}) \right\}^{y_{it}}} \right]^{w_i}$$

$$= \log \prod_{i=1}^{n} \left\{ \frac{(\sum_t y_{it})!}{\prod_{t=1}^{n_i} y_{it}!} \prod_{t=1}^{n_i} p_{it}^{y_{it}} \right\}^{w_i}$$

$$= \sum_{i=1}^{n} w_i \left\{ \log \Gamma \left(\sum_{t=1}^{n_i} y_{it} + 1 \right) - \sum_{t=1}^{n_i} \log \Gamma(y_{it} + 1) + \sum_{t=1}^{n_i} y_{it} \log p_{it} \right\}$$

$$p_{it} = e^{\mathbf{x}_{it}\boldsymbol{\beta}} \Big/ \sum_{\ell} e^{\mathbf{x}_{i\ell}\boldsymbol{\beta}}$$

References

Cameron, A. C. and P. K. Trivedi. 1998. *Regression Analysis of Count Data.* New York: Cambridge University Press.

Greene, W. H. 2003. *Econometric Analysis.* 5th ed. Upper Saddle River, NJ: Prentice–Hall.

Hardin, J. and J. Hilbe. 2001. *Generalized Linear Models and Extensions.* College Station, TX: Stata Press.

Hausman, J., B. H. Hall, and Z. Griliches. 1984. Econometric models for count data with an application to the patents–R & D relationship. *Econometrica* 52: 909–938.

Liang, K.-Y. and S. L. Zeger. 1986. Longitudinal data analysis using generalized linear models. *Biometrika* 73: 13–22.

McCullagh, P. and J. A. Nelder. 1989. *Generalized Linear Models.* 2d ed. London: Chapman & Hall.

Also See

Complementary:	[XT] **quadchk**, [XT] **xtdata**, [XT] **xtdes**, [XT] **xtsum**, [XT] **xttab**, [R] **adjust**, [R] **constraint**, [R] **lincom**, [R] **mfx**, [R] **nlcom**, [R] **predict**, [R] **predictnl**, [R] **test**, [R] **testnl**, [R] **vce**
Related:	[XT] **xtgee**, [XT] **xtnbreg**, [R] **poisson**
Background:	[U] **16.5 Accessing coefficients and standard errors**, [U] **23 Estimation and post-estimation commands**, [U] **23.14 Obtaining robust variance estimates**, [XT] **xt**

Title

xtprobit — Random-effects and population-averaged probit models

Syntax

Random-effects model

> xtprobit *depvar* [*varlist*] [*weight*] [if *exp*] [in *range*] [, re i(*varname*) quad(*#*)
>
> noconstant noskip level(*#*) offset(*varname*) nolog *maximize_options*]

Population-averaged model

> xtprobit *depvar* [*varlist*] [*weight*] [if *exp*] [in *range*] , pa [i(*varname*) robust
>
> noconstant level(*#*) offset(*varname*) nolog *xtgee_options* *maximize_options*]

by ... : may be used with xtprobit; see [R] **by**.

iweights, fweights, and pweights are allowed for the population-averaged model and iweights are allowed in the random-effects model; see [U] **14.1.6 weight**. Note that weights must be constant within panels.

xtprobit shares the features of all estimation commands; see [U] **23 Estimation and post-estimation commands**.

Syntax for predict

Random-effects model

> predict [*type*] *newvarname* [if *exp*] [in *range*] [, [xb | pu0 | stdp]
>
> nooffset]

Population-averaged model

> predict [*type*] *newvarname* [if *exp*] [in *range*] [, [mu | rate | xb | stdp]
>
> nooffset]

These statistics are available both in and out of sample; type predict ... if e(sample) ... if wanted only for the estimation sample.

Description

xtprobit fits random-effects and population-averaged probit models. There is no command for a conditional fixed-effects model, as there does not exist a sufficient statistic allowing the fixed effects to be conditioned out of the likelihood. Unconditional fixed-effects probit models may be fitted with the probit command with indicator variables for the panels. The appropriate indicator variables can be generated using tabulate or xi. However, unconditional fixed-effects estimates are biased.

Note: xtprobit, re, the default, is slow since it is calculated by quadrature; see *Methods and Formulas*. Computation time is roughly proportional to the number of points used for the quadrature. The default is quad(12). Simulations indicate that increasing it does not appreciably change the estimates for the coefficients or their standard errors. See [XT] **quadchk**.

By default, the population-averaged model is an equal-correlation model; xtprobit assumes the within-group correlation structure corr(exchangeable). See [XT] **xtgee** for information on how to fit other population-averaged models.

See [R] **logistic** for a list of related estimation commands.

Options

re requests the random-effects estimator. re is the default if neither re nor pa is specified.

pa requests the population-averaged estimator.

i(*varname*) specifies the variable name that contains the unit to which the observation belongs. You can specify the i() option the first time you estimate, or you can use the iis command to set i() beforehand. Note that it is not necessary to specify i() if the data have been previously tsset, or if iis has been previously specified—in these cases, the group variable is taken from the previous setting. See [XT] **xt**.

quad(*#*) specifies the number of points to use in the quadrature approximation of the integral. This option is relevant only if you are fitting a random-effects model; if you specify quad(*#*) with pa, the quad(*#*) is ignored.

The default is quad(12). The number specified must be an integer between 4 and 30, and must be no greater than the number of observations.

noconstant suppresses the constant term (intercept) in the model.

noskip specifies that a full maximum-likelihood model with only a constant for the regression equation be fitted. This model is not displayed, but is used as the base model to compute a likelihood-ratio test for the model test statistic displayed in the estimation header. By default, the overall model test statistic is an asymptotically equivalent Wald test of all the parameters in the regression equation being zero (except the constant). For many models, this option can substantially increase estimation time.

level(*#*) specifies the confidence level, in percent, for confidence intervals. The default is level(95) or as set by set level; see [U] **23.6 Specifying the width of confidence intervals**.

offset(*varname*) specifies that *varname* is to be included in the model with its coefficient constrained to be 1.

robust specifies that the Huber/White/sandwich estimator of variance is to be used in place of the IRLS variance estimator; see [XT] **xtgee**. This alternative produces valid standard errors even if the correlations within group are not as hypothesized by the specified correlation structure. It does, however, require that the model correctly specifies the mean. As such, the resulting standard errors are labeled "semi-robust" instead of "robust". Note that although there is no cluster() option, results are as if there were a cluster() option and you specified clustering on i().

nolog suppresses the iteration log.

xtgee_options specifies any other options allowed by xtgee for family(binomial) link(probit), such as corr(); see [XT] **xtgee**.

maximize_options control the maximization process; see [R] **maximize**. Use the `trace` option to view parameter convergence. Use the `ltol(#)` option to relax the convergence criterion; the default is 1e−6 during specification searches.

Options for predict

`xb` calculates the linear prediction. This is the default for the random-effects model.

`pu0` calculates the probability of a positive outcome, assuming that the random effect for that observation's panel is zero ($\nu = 0$). Note that this may not be similar to the proportion of observed outcomes in the group.

`stdp` calculates the standard error of the linear prediction.

`mu` and `rate` both calculate the predicted probability of *depvar*. `mu` takes into account the `offset()`, and `rate` ignores those adjustments. `mu` and `rate` are equivalent if you did not specify `offset()`. `mu` is the default for the population-averaged model.

`nooffset` is relevant only if you specified `offset(`*varname*`)` for `xtprobit`. It modifies the calculations made by `predict` so that they ignore the offset variable; the linear prediction is treated as $\mathbf{x}_{it}\boldsymbol{\beta}$ rather than $\mathbf{x}_{it}\boldsymbol{\beta} + \text{offset}_{it}$.

Remarks

`xtprobit` is a convenience command if you want the population-averaged model. Typing

        . xtprobit ..., pa ...

is equivalent to typing

        . xtgee ..., ... family(binomial) link(probit) corr(exchangeable)

Thus, also see [XT] **xtgee** for information about `xtprobit`.

By default, or when `re` is specified, `xtprobit` fits via maximum-likelihood the random-effects model

$$\Pr(y_{it} \neq 0|\mathbf{x}_{it}) = \Phi(\mathbf{x}_{it}\boldsymbol{\beta} + \nu_i)$$

for $i = 1, \ldots, n$ panels, $t = 1, \ldots, n_i$, ν_i are iid $N(0, \sigma_\nu^2)$, and Φ is the standard normal cumulative distribution function.

Underlying this model is the variance components model

$$y_{it} \neq 0 \iff \mathbf{x}_{it}\boldsymbol{\beta} + \nu_i + \epsilon_{it} > 0$$

where ϵ_{it} are iid Gaussian distributed with mean zero and variance $\sigma_\epsilon^2 = 1$, independently of ν_i.

▷ Example

You are studying unionization of women in the United States and are using the `union` dataset; see [XT] **xt**. You wish to fit a random-effects model of union membership:

(Continued on next page)

```
. use http://www.stata-press.com/data/r8/union
(NLS Women 14-24 in 1968)

. xtprobit union age grade not_smsa south southXt, i(id) nolog
```

Random-effects probit regression   Number of obs   =   26200
Group variable (i): idcode   Number of groups   =   4434

Random effects u_i ~ Gaussian   Obs per group: min =   1
                                                                   avg =   5.9
                                                                   max =   12

                                                    Wald chi2(5)   =   218.90
Log likelihood = -10561.065   Prob > chi2   =   0.0000

union	Coef.	Std. Err.	z	P>|z|	[95% Conf. Interval]	
age	.0044483	.0025027	1.78	0.076	-.000457	.0093535
grade	.0482482	.0100413	4.80	0.000	.0285677	.0679287
not_smsa	-.1370699	.0462961	-2.96	0.003	-.2278087	-.0463312
south	-.6305824	.0614827	-10.26	0.000	-.7510863	-.5100785
southXt	.0131853	.0043819	3.01	0.003	.004597	.0217737
_cons	-1.846838	.1458222	-12.67	0.000	-2.132644	-1.561032
/lnsig2u	.5612193	.0431875			.4765733	.6458653
sigma_u	1.323937	.0285888			1.269073	1.381172
rho	.6367346	.0099894			.6169384	.6560781

Likelihood-ratio test of rho=0: chibar2(01) = 5972.49 Prob >= chibar2 = 0.000

The output includes the additional panel-level variance component. This is parameterized as the log of the variance $\ln(\sigma_\nu^2)$ (labeled lnsig2u in the output). The standard deviation σ_ν is also included in the output labeled sigma_u together with ρ (labeled rho),

$$\rho = \frac{\sigma_\nu^2}{\sigma_\nu^2 + 1}$$

which is the proportion of the total variance contributed by the panel-level variance component.

When rho is zero, the panel-level variance component is unimportant and the panel estimator is not different from the pooled estimator. A likelihood-ratio test of this is included at the bottom of the output. This test formally compares the pooled estimator (probit) with the panel estimator.

◁

❏ Technical Note

The random-effects model is calculated using quadrature. As the panel sizes (or ρ) increase, the quadrature approximation becomes less accurate. We can use the quadchk command to see if changing the number of quadrature points affects the results. If the results do change, then the quadrature approximation is not accurate, and the results of the model should not be interpreted.

(Continued on next page)

```
. quadchk, nooutput
Refitting model quad() =  8
Refitting model quad() = 16
```

Quadrature check

	Fitted quadrature 12 points	Comparison quadrature 8 points	Comparison quadrature 16 points	
Log likelihood	-10561.065	-10574.78	-10555.853	
		-13.714764	5.2126898	Difference
		.00129862	-.00049358	Relative difference
union: age	.00444829	.00478943	.00451117	
		.00034115	.00006288	Difference
		.07669143	.01413662	Relative difference
union: grade	.04824822	.05629525	.04411081	
		.00804704	-.00413741	Difference
		.16678412	-.0857525	Relative difference
union: not_smsa	-.13706993	-.1314541	-.14109796	
		.00561584	-.00402803	Difference
		-.04097061	.02938665	Relative difference
union: south	-.63058241	-.62309654	-.64546968	
		.00748587	-.01488727	Difference
		-.01187136	.02360876	Relative difference
union: southXt	.01318534	.01194434	.01341723	
		-.001241	.00023189	Difference
		-.09411977	.01758658	Relative difference
union: _cons	-1.8468379	-1.9306422	-1.8066853	
		-.08380426	.0401526	Difference
		.04537716	-.02174127	Relative difference
lnsig2u: _cons	.56121927	.49078989	.58080961	
		-.07042938	.01959034	Difference
		-.12549352	.03490674	Relative difference

Note that the results obtained for 12 quadrature points were closer to the results using 18 points than to the results using 6 points. However, since the convergence point seems to be sensitive to the number of quadrature points, we should not use this output. Since there is no alternative method for calculating the random-effects model in Stata, we may either fit a different model or use a different command. We should not use the output of a random-effects specification when there is evidence that the numeric technique for calculating the model is not stable (as shown by quadchk).

A subjective rule of thumb is that the relative differences in the coefficients should not change by more than 1% if the quadrature technique is stable. See [XT] **quadchk** for details. The important point to remember is that when the quadrature technique is not stable, you cannot merely increase the number of quadrature points to fix the problem.

❑

▷ Example

As an alternative to the random-effects specification, we can fit an equal-correlation probit model:

```
. xtprobit union age grade not_smsa south southXt, i(id) pa
Iteration 1: tolerance = .04796083
Iteration 2: tolerance = .00352657
Iteration 3: tolerance = .00017886
Iteration 4: tolerance = 8.654e-06
Iteration 5: tolerance = 4.150e-07
```

```
GEE population-averaged model              Number of obs      =       26200
Group variable:                   idcode   Number of groups   =        4434
Link:                             probit   Obs per group: min =           1
Family:                         binomial                  avg =         5.9
Correlation:                 exchangeable                  max =          12
                                           Wald chi2(5)       =      241.66
Scale parameter:                       1   Prob > chi2        =      0.0000
```

union	Coef.	Std. Err.	z	P>|z|	[95% Conf. Interval]	
age	.0031597	.0014678	2.15	0.031	.0002829	.0060366
grade	.0329992	.0062334	5.29	0.000	.020782	.0452163
not_smsa	-.0721799	.0275189	-2.62	0.009	-.1261159	-.0182439
south	-.409029	.0372213	-10.99	0.000	-.4819815	-.3360765
southXt	.0081828	.002545	3.22	0.001	.0031946	.0131709
_cons	-1.184799	.0890117	-13.31	0.000	-1.359259	-1.01034

◁

▷ Example

In [R] **probit**, we showed the above results and compared them with probit, robust cluster(). xtprobit with the pa option allows a robust option (the random-effects estimator does not allow the robust specification), so we can obtain the population-averaged probit estimator with the robust variance calculation by typing

```
. xtprobit union age grade not_smsa south southXt, i(id) pa robust nolog
```

```
GEE population-averaged model              Number of obs      =       26200
Group variable:                   idcode   Number of groups   =        4434
Link:                             probit   Obs per group: min =           1
Family:                         binomial                  avg =         5.9
Correlation:                 exchangeable                  max =          12
                                           Wald chi2(5)       =      154.00
Scale parameter:                       1   Prob > chi2        =      0.0000
```

(standard errors adjusted for clustering on idcode)

union	Coef.	Semi-robust Std. Err.	z	P>|z|	[95% Conf. Interval]	
age	.0031597	.0022027	1.43	0.151	-.0011575	.007477
grade	.0329992	.0076631	4.31	0.000	.0179797	.0480186
not_smsa	-.0721799	.0348772	-2.07	0.038	-.140538	-.0038218
south	-.409029	.0482545	-8.48	0.000	-.5036061	-.3144519
southXt	.0081828	.0037108	2.21	0.027	.0009097	.0154558
_cons	-1.184799	.116457	-10.17	0.000	-1.413051	-.9565479

These standard errors are similar to those shown for probit, robust cluster() in [R] **probit**.

◁

▷ Example

In a previous example, we showed how `quadchk` indicated that the quadrature technique was numerically unstable. Here, we present an example where the quadrature is stable.

In this example, we have (synthetic) data on whether workers complain to managers at a fast-food restaurant. The covariates are `age` (in years of the worker), `grade` (years of schooling completed by worker), `south` (equal to 1 if the restaurant is located in the South), `tenure` (the number of years spent on the job by the worker), `gender` (of worker), `race` (of work), `income` (in thousands of dollars by the restaurant), `genderm` (gender of manager), `burger` (equal to 1 if the restaurant specializes in hamburgers), and `chicken` (equal to 1 if the restaurant specializes in chicken). The model is given by

```
. use http://www.stata-press.com/data/chicken
. xtprobit complain age grade south tenure gender race income genderm
> burger chicken, i(person) nolog
```

Random-effects probit regression	Number of obs	=	5952
Group variable (i): person	Number of groups	=	1076
Random effects u_i ~ Gaussian	Obs per group: min =		3
	avg =		5.5
	max =		8
	Wald chi2(10)	=	65.03
Log likelihood = -2574.115	Prob > chi2	=	0.0000

complain	Coef.	Std. Err.	z	P>|z|	[95% Conf. Interval]	
age	-.0003157	.0762518	-0.00	0.997	-.1497664	.1491351
grade	-.0411126	.0647727	-0.63	0.526	-.1680648	.0858396
south	-.0346366	.0723753	-0.48	0.632	-.1764895	.1072163
tenure	-.3836063	.0550447	-6.97	0.000	-.4914919	-.2757206
gender	.0667994	.0734595	0.91	0.363	-.0771786	.2107775
race	.0834963	.055773	1.50	0.134	-.0258168	.1928094
income	-.2111629	.0730126	-2.89	0.004	-.354265	-.0680607
genderm	.1306497	.0557133	2.35	0.019	.0214535	.2398458
burger	-.0616544	.0729739	-0.84	0.398	-.2046806	.0813718
chicken	.0635842	.0557645	1.14	0.254	-.0457122	.1728806
_cons	-1.123845	.0330159	-34.04	0.000	-1.188555	-1.059136
/lnsig2u	-1.030313	.1292422			-1.283623	-.7770027
sigma_u	.5974072	.0386051			.5263382	.6780723
rho	.2630235	.0250526			.2169342	.3149662

Likelihood-ratio test of rho=0: chibar2(01) =   166.88 Prob >= chibar2 = 0.000

Again, we would like to check the stability of the quadrature technique of the model before interpreting the results. Given the estimate of ρ and the small size of the panels (between 3 and 8), we should find that the quadrature technique is numerically stable.

(Continued on next page)

```
. quadchk, nooutput
```

Refitting model quad() = 8
Refitting model quad() = 16

	Fitted quadrature 12 points	Comparison quadrature 8 points	Comparison quadrature 16 points	
		Quadrature check		
Log likelihood	-2574.115	-2574.1293	-2574.1164	
		-.01424246	-.00132578	Difference
		5.533e-06	5.150e-07	Relative difference
complain: age	-.00031569	-.00013111	-.00031858	
		.00018459	-2.891e-06	Difference
		-.58470325	.00915831	Relative difference
complain: grade	-.04111263	-.04100666	-.04111079	
		.00010597	1.839e-06	Difference
		-.0025776	-.00004474	Relative difference
complain: south	-.03463663	-.03469524	-.03462929	
		-.00005861	7.341e-06	Difference
		.0016922	-.00021193	Relative difference
complain: tenure	-.38360629	-.38351811	-.38360047	
		.00008818	5.820e-06	Difference
		-.00022987	-.00001517	Relative difference
complain: gender	.06679944	.06655282	.0668029	
		-.00024662	3.455e-06	Difference
		-.003692	.00005172	Relative difference
complain: race	.0834963	.08340258	.08349099	
		-.00009373	-5.311e-06	Difference
		-.00112252	-.00006361	Relative difference
complain: income	-.21116286	-.21100203	-.21115631	
		.00016083	6.556e-06	Difference
		-.00076164	-.00003105	Relative difference
complain: genderm	.13064966	.1305605	.13064386	
		-.00008916	-5.804e-06	Difference
		-.00068243	-.00004442	Relative difference
complain: burger	-.06165444	-.06168062	-.06164754	
		-.00002618	6.903e-06	Difference
		.0004246	-.00011195	Relative difference
complain: chicken	.0635842	.06359848	.06358665	
		.00001428	2.452e-06	Difference
		.00022456	.00003856	Relative difference
complain: _cons	-1.1238455	-1.1237932	-1.1238278	
		.00005231	.00001769	Difference
		-.00004654	-.00001574	Relative difference
lnsig2u: _cons	-1.0303127	-1.0317132	-1.0304345	
		-.00140056	-.0001218	Difference
		.00135936	.00011821	Relative difference

The relative differences are all very small between the default 12 quadrature points and the result with 16 points. We have only one coefficient that has a large relative difference between the default 12 quadrature points and 8 quadrature points. In looking again at the absolute differences, we see also that the absolute differences between 12 and 16 quadrature points were also small.

We conclude that the quadrature technique is stable. We may wish to rerun the above model with quad(16) or even higher (but we do not have to since the results will not significantly differ) and interpret those results for our presentation.

◁

Saved Results

xtprobit, re saves in e():

Scalars

e(N)	number of observations	e(g_avg)	average group size
e(N_g)	number of groups	e(chi2)	χ^2
e(df_m)	model degrees of freedom	e(chi2_c)	χ^2 for comparison test
e(ll)	log likelihood	e(rho)	ρ
e(ll_0)	log likelihood, constant-only model	e(sigma_u)	panel-level standard deviation
e(ll_c)	log likelihood, comparison model	e(N_cd)	number of completely determined obs.
e(g_max)	largest group size	e(n_quad)	number of quadrature points
e(g_min)	smallest group size		

Macros

e(cmd)	xtprobit	e(chi2type)	Wald or LR; type of model χ^2 test
e(depvar)	name of dependent variable	e(chi2_ct)	Wald or LR; type of model χ^2 test corresponding to e(chi2_c)
e(title)	title in estimation output		
e(ivar)	variable denoting groups	e(distrib)	Gaussian; the distribution of the random effect
e(wtype)	weight type		
e(wexp)	weight expression	e(crittype)	optimization criterion
e(offset)	offset	e(predict)	program used to implement predict

Matrices

e(b)	coefficient vector	e(V)	variance–covariance matrix of the estimators

Functions

e(sample)	marks estimation sample

(*Continued on next page*)

`xtprobit, pa` saves in e():

Scalars

e(N)	number of observations	e(deviance)	deviance
e(N_g)	number of groups	e(chi2_dev)	χ^2 test of deviance
e(df_m)	model degrees of freedom	e(dispers)	deviance dispersion
e(g_max)	largest group size	e(chi2_dis)	χ^2 test of deviance dispersion
e(g_min)	smallest group size	e(tol)	target tolerance
e(g_avg)	average group size	e(dif)	achieved tolerance
e(chi2)	χ^2	e(phi)	scale parameter
e(df_pear)	degrees of freedom for Pearson χ^2		

Macros

e(cmd)	xtgee	e(ivar)	variable denoting groups
e(cmd2)	xtprobit	e(vcetype)	covariance estimation method
e(depvar)	name of dependent variable	e(chi2type)	Wald; type of model χ^2 test
e(family)	binomial	e(offset)	offset
e(link)	probit; link function	e(crittype)	optimization criterion
e(corr)	correlation structure	e(predict)	program used to implement predict
e(scale)	x2, dev, phi, or #; scale parameter		

Matrices

e(b)	coefficient vector	e(R)	estimated working correlation matrix
e(V)	variance–covariance matrix of the estimators		

Functions

e(sample)	marks estimation sample

Methods and Formulas

xtprobit is implemented as an ado-file.

xtprobit reports the population-averaged results obtained by using `xtgee, family(binomial) link(probit)` to obtain estimates.

Assuming a normal distribution, $N(0, \sigma_\nu^2)$, for the random effects ν_i, we have that

$$\Pr(y_{i1}, \ldots, y_{in_i} | \mathbf{x}_{i1}, \ldots, \mathbf{x}_{in_i}) = \int_{-\infty}^{\infty} \frac{e^{-\nu_i^2/2\sigma_\nu^2}}{\sqrt{2\pi}\sigma_\nu} \left\{ \prod_{t=1}^{n_i} F(y_{it}, \mathbf{x}_{it}\boldsymbol{\beta} + \nu_i) \right\} d\nu_i$$

where

$$F(y, z) = \begin{cases} \Phi(z) & \text{if } y \neq 0 \\ 1 - \Phi(z) & \text{otherwise} \end{cases}$$

(where Φ is the cumulative normal distribution), and we can approximate the integral with M-point Gauss–Hermite quadrature

$$\int_{-\infty}^{\infty} e^{-x^2} g(x) dx \approx \sum_{m=1}^{M} w_m^* g(a_m^*)$$

where the w_m^* denote the quadrature weights and the a_m^* denote the quadrature abscissas. The log-likelihood L, where $\rho = \sigma_\nu^2/(\sigma_\nu^2 + 1)$, is then calculated using the quadrature

$$L = \sum_{i=1}^{n} w_i \log \left\{ \Pr(y_{i1}, \ldots, y_{in_i} | \mathbf{x}_{i1}, \ldots, \mathbf{x}_{in_i}) \right\}$$

$$\approx \sum_{i=1}^{n} w_i \log \left[\frac{1}{\sqrt{\pi}} \sum_{m=1}^{M} w_m^* \prod_{t=1}^{n_i} F \left\{ y_{it}, \mathbf{x}_{it}\boldsymbol{\beta} + a_m^* \left(\frac{2\rho}{1-\rho} \right)^{1/2} \right\} \right]$$

where w_i is the user-specified weight for panel i; if no weights are specified, $w_i = 1$.

The quadrature formula requires that the integrated function be well-approximated by a polynomial. As the number of time periods becomes large (as panel size gets large),

$$\prod_{t=1}^{n_i} F(y_{it}, \mathbf{x}_{it}\boldsymbol{\beta} + \nu_i)$$

is no longer well-approximated by a polynomial. As a general rule of thumb, you should use this quadrature approach only for small to moderate panel sizes (based on simulations, 50 is a reasonably safe upper bound). However, if the data really come from random-effects probit and ρ is not too large (less than, say, .3), then the panel size could be 500 and the quadrature approximation would still be fine. If the data are not random-effects probit or ρ is large (bigger than, say, .7), then the quadrature approximation may be poor for panel sizes larger than 10. The `quadchk` command should be used to investigate the applicability of the numeric technique used in this command.

References

Conway, M. R. 1990. A random effects model for binary data. *Biometrics* 46: 317–328.

Frechette, G. R. 2001. sg158: Random-effects ordered probit. *Stata Technical Bulletin* 59: 23–27. Reprinted in *Stata Technical Bulletin Reprints*, vol. 10, pp. 261–266.

——. 2001. sg158.1: Update to random-effects probit. *Stata Technical Bulletin* 61: 12. Reprinted in *Stata Technical Bulletin Reprints*, vol. 10, pp. 266–267.

Guilkey, D. K. and J. L. Murphy. 1993. Estimation and testing in the random effects probit model. *Journal of Econometrics* 59: 301–317.

Liang, K.-Y. and S. L. Zeger. 1986. Longitudinal data analysis using generalized linear models. *Biometrika* 73: 13–22.

Neuhaus, J. M. 1992. Statistical methods for longitudinal and clustered designs with binary responses. *Statistical Methods in Medical Research* 1: 249–273.

Neuhaus, J. M., J. D. Kalbfleisch, and W. W. Hauck. 1991. A comparison of cluster-specific and population-averaged approaches for analyzing correlated binary data. *International Statistical Review* 59: 25–35.

Pendergast, J. F., S. J. Gange, M. A. Newton, M. J. Lindstrom, M. Palta, and M. R. Fisher. 1996. A survey of methods for analyzing clustered binary response data. *International Statistical Review* 64: 89–118.

Also See

Complementary:	[XT] **quadchk**, [XT] **xtdata**, [XT] **xtdes**, [XT] **xtsum**, [XT] **xttab**,
	[R] **adjust**, [R] **lincom**, [R] **mfx**, [R] **nlcom**, [R] **predict**, [R] **predictnl**,
	[R] **test**, [R] **testnl**, [R] **vce**
Related:	[XT] **xtcloglog**, [XT] **xtgee**, [XT] **xtlogit**,
	[R] **probit**
Background:	[U] **16.5 Accessing coefficients and standard errors**,
	[U] **23 Estimation and post-estimation commands**,
	[U] **23.14 Obtaining robust variance estimates**,
	[XT] **xt**

Title

> **xtrchh** — Hildreth–Houck random-coefficients model

Syntax

> xtrchh *depvar varlist* [if *exp*] [in *range*] [, i(*varname*) t(*varname*) level(*#*)
>
> offset(*varname*) noconstant *maximize_options*]

by ... : may be used with xtrchh; see [R] **by**.

xtrchh shares the features of all estimation commands; see [U] **23 Estimation and post-estimation commands**.

Syntax for predict

> predict [*type*] *newvarname* [if *exp*] [in *range*] [, [xb | stdp] nooffset]

These statistics are available both in and out of sample; type predict ... if e(sample) ... if wanted only for the estimation sample.

Description

xtrchh fits the Hildreth–Houck random-coefficients linear regression model.

Options

i(*varname*) specifies the variable that contains the unit to which the observation belongs. You can specify the i() option the first time you estimate, or you can use the iis command to set i() beforehand. Note that it is not necessary to specify i() if the data have been previously tsset, or if iis has been previously specified—in these cases, the group variable is taken from the previous setting. See [XT] **xt**.

t(*varname*) specifies the variable that contains the time at which the observation was made. You can specify the t() option the first time you estimate, or you can use the tis command to set t() beforehand. Note that it is not necessary to specify t() if the data have been previously tsset, or if tis has been previously specified—in these cases, the group variable is taken from the previous setting. See [XT] **xt**.

level(*#*) specifies the confidence level, in percent, for confidence intervals. The default is level(95) or as set by set level; see [U] **23.6 Specifying the width of confidence intervals**.

offset(*varname*) specifies that *varname* is to be included in the model with its coefficient constrained to be 1.

noconstant suppresses the constant term (intercept) in the regression.

maximize_options control the maximization process; see [R] **maximize**. Use the trace option to view parameter convergence. Use the ltol(*#*) option to relax the convergence criterion; the default is 1e−6 during specification searches.

Options for predict

xb, the default, calculates the linear prediction.

stdp calculates the standard error of the linear prediction.

nooffset is relevant only if you specified offset(*varname*) for xtrchh. It modifies the calculations made by predict so that they ignore the offset variable; the linear prediction is treated as $\mathbf{x}_{it}\mathbf{b}$ rather than $\mathbf{x}_{it}\mathbf{b} + \text{offset}_{it}$.

Remarks

In random-coefficients models, we wish to treat the parameter vector as a realization (in each panel) of a stochastic process.

Interested readers should see Greene (2003) for information on this and other panel-data models.

▷ Example

Greene (2003, 329 and *http://www.prenhall.com/greene*) reprints data from a classic study of investment demand by Grunfeld and Griliches (1960). In [XT] **xtgls**, we use this dataset to illustrate many of the possible models that may be fitted with the xtgls command. While the models included in the xtgls command offer considerable flexibility, they all assume that there is no parameter variation across firms (the cross-sectional units).

To take a first look at the assumption of parameter constancy, we should reshape our data so that we may fit a simultaneous equation model using sureg; see [R] **sureg**. Since there are only five panels here, it is not too difficult.

```
. use http://www.stata-press.com/data/r8/invest2

. reshape wide invest market stock, i(time) j(company)
(note: j = 1 2 3 4 5)
```

Data	long	->	wide
Number of obs.	100	->	20
Number of variables	5	->	16
j variable (5 values)	company	->	(dropped)
xij variables:			
	invest	->	invest1 invest2 ... invest5
	market	->	market1 market2 ... market5
	stock	->	stock1 stock2 ... stock5

```
. sureg (invest1 market1 stock1) (invest2 market2 stock2) (invest3 market3 stock3)
> (invest4 market4 stock4) (invest5 market5 stock5)

Seemingly unrelated regression
```

Equation	Obs	Parms	RMSE	"R-sq"	chi2	P
invest1	20	2	84.94729	0.9207	261.32	0.0000
invest2	20	2	12.36322	0.9119	207.21	0.0000
invest3	20	2	26.46612	0.6876	46.88	0.0000
invest4	20	2	9.742303	0.7264	59.15	0.0000
invest5	20	2	95.85484	0.4220	14.97	0.0006

	Coef.	Std. Err.	z	P>|z|	[95% Conf.	Interval]
invest1						
market1	.120493	.0216291	5.57	0.000	.0781007	.1628853
stock1	.3827462	.032768	11.68	0.000	.318522	.4469703
_cons	-162.3641	89.45922	-1.81	0.070	-337.7009	12.97279
invest2						
market2	.0695456	.0168975	4.12	0.000	.0364271	.1026641
stock2	.3085445	.0258635	11.93	0.000	.2578529	.3592362
_cons	.5043112	11.51283	0.04	0.965	-22.06042	23.06904
invest3						
market3	.0372914	.0122631	3.04	0.002	.0132561	.0613268
stock3	.130783	.0220497	5.93	0.000	.0875663	.1739997
_cons	-22.43892	25.51859	-0.88	0.379	-72.45443	27.57659
invest4						
market4	.0570091	.0113623	5.02	0.000	.0347395	.0792788
stock4	.0415065	.0412016	1.01	0.314	-.0392472	.1222602
_cons	1.088878	6.258805	0.17	0.862	-11.17815	13.35591
invest5						
market5	.1014782	.0547837	1.85	0.064	-.0058958	.2088523
stock5	.3999914	.1277946	3.13	0.002	.1495186	.6504642
_cons	85.42324	111.8774	0.76	0.445	-133.8525	304.6989

Here, we instead fit a random-coefficients model:

```
. use http://www.stata-press.com/data/r8/invest2, clear

. xtrchh invest market stock, i(company) t(time)
```

Hildreth-Houck random-coefficients regression Number of obs   =   100
Group variable (i): company   Number of groups =   5

                                               Obs per group: min =   20
                                                           avg =   20.0
                                                           max =   20

                                               Wald chi2(2)   =   17.55
                                               Prob > chi2   =   0.0002

invest	Coef.	Std. Err.	z	P>|z|	[95% Conf.	Interval]
market	.0807646	.0250829	3.22	0.001	.0316031	.1299261
stock	.2839885	.0677899	4.19	0.000	.1511229	.4168542
_cons	-23.58361	34.55547	-0.68	0.495	-91.31108	44.14386

Test of parameter constancy: chi2(12) =   603.99   Prob > chi2 = 0.0000

Just as a subjective examination of the results of our simultaneous-equation model does not support the assumption of parameter constancy, the test included with the random-coefficients model also indicates that assumption is not valid for these data. With large panel datasets, obviously we would not want to take the time to look at a simultaneous-equations model (aside from the fact that our doing so was very subjective).

◁

Saved Results

xtrchh saves in e():

Scalars

e(N)	number of observations	e(g_avg)	average group size
e(N_g)	number of groups	e(chi2)	χ^2
e(df_m)	model degrees of freedom	e(chi2_c)	χ^2 for comparison test
e(g_max)	largest group size	e(df_chi2)	degrees of freedom for model χ^2
e(g_min)	smallest group size		

Macros

e(cmd)	xtrchh	e(chi2type)	Wald; type of model χ^2 test
e(depvar)	name of dependent variable	e(offset)	offset
e(title)	title in estimation output	e(predict)	program used to implement predict
e(ivar)	variable denoting groups		

Matrices

e(b)	coefficient vector	e(V)	variance–covariance matrix of the estimators
e(Sigma)	$\widehat{\Sigma}$ matrix		

Functions

e(sample)	marks estimation sample

Methods and Formulas

xtrchh is implemented as an ado-file.

In a random-coefficients model, the parameter heterogeneity is treated as stochastic variation. Assume that we write

$$\mathbf{y}_i = \mathbf{X}_i \boldsymbol{\beta}_i + \boldsymbol{\epsilon}_i$$

where $i = 1, \ldots, m$, and $\boldsymbol{\beta}_i$ is the coefficient vector ($k \times 1$) for the ith cross-sectional unit such that

$$\boldsymbol{\beta}_i = \boldsymbol{\beta} + \boldsymbol{\nu}_i \qquad E(\boldsymbol{\nu}_i) = \mathbf{0} \qquad E(\boldsymbol{\nu}_i \boldsymbol{\nu}_i') = \boldsymbol{\Sigma}$$

Our goal is to find $\widehat{\boldsymbol{\beta}}$ and $\widehat{\boldsymbol{\Sigma}}$.

The derivation of the estimator assumes that the cross-sectional specific coefficient vector $\boldsymbol{\beta}_i$ is the outcome of a random process with mean vector $\boldsymbol{\beta}$ and covariance matrix $\boldsymbol{\Sigma}$,

$$\mathbf{y}_i = \mathbf{X}_i \boldsymbol{\beta}_i + \boldsymbol{\epsilon}_i = \mathbf{X}_i (\boldsymbol{\beta} + \boldsymbol{\nu}_i) + \boldsymbol{\epsilon}_i = \mathbf{X}_i \boldsymbol{\beta} + (\mathbf{X}_i \boldsymbol{\nu}_i + \boldsymbol{\epsilon}_i) = \mathbf{X}_i \boldsymbol{\beta} + \boldsymbol{\omega}_i$$

where $E(\boldsymbol{\omega}_i) = \mathbf{0}$ and

$$E(\boldsymbol{\omega}_i \boldsymbol{\omega}_i') = E\left\{ (\mathbf{X}_i \boldsymbol{\nu}_i + \boldsymbol{\epsilon}_i)(\mathbf{X}_i \boldsymbol{\nu}_i + \boldsymbol{\epsilon}_i)' \right\} = E(\boldsymbol{\epsilon}_i \boldsymbol{\epsilon}_i') + \mathbf{X}_i E(\boldsymbol{\nu}_i \boldsymbol{\nu}_i') \mathbf{X}_i' = \sigma_i^2 \mathbf{I} + \mathbf{X}_i \boldsymbol{\Sigma} \mathbf{X}_i' = \boldsymbol{\Pi}_i$$

The covariance matrix for the panel-specific coefficient estimator $\boldsymbol{\beta}_i$ can then be written as

$$\mathbf{V}_i + \boldsymbol{\Sigma} = (\mathbf{X}_i' \mathbf{X}_i)^{-1} \mathbf{X}_i' \boldsymbol{\Pi}_i \mathbf{X}_i (\mathbf{X}_i' \mathbf{X}_i)^{-1} \qquad \text{where} \qquad \mathbf{V}_i = \sigma_i^2 (\mathbf{X}_i' \mathbf{X}_i)^{-1}$$

We may then compute a weighted average of the panel-specific coefficient estimates as

$$\widehat{\boldsymbol{\beta}} = \sum_{i=1}^{m} \mathbf{W}_i \boldsymbol{\beta}_i \qquad \text{where} \qquad \mathbf{W}_i = \left\{ \sum_{i=1}^{m} (\boldsymbol{\Sigma} + \mathbf{V}_i)^{-1} \right\}^{-1} (\boldsymbol{\Sigma} + \mathbf{V}_i)^{-1}$$

such that the resulting GLS estimator is a matrix-weighted average of the panel-specific (OLS) estimators.

To calculate the above estimator $\widehat{\beta}$ for the unknown Σ and $\mathbf{V}_i$ parameters, we use the two-step approach suggested by Swamy (1971):

$$\widehat{\beta}_i = \text{OLS panel-specific estimator}$$

$$\widehat{\mathbf{V}}_i = \frac{\widehat{\epsilon}_i'\widehat{\epsilon}_i}{n_i - k}$$

$$\overline{\beta} = \frac{1}{m}\sum_{i=1}^{m}\widehat{\beta}_i$$

$$\widehat{\Sigma} = \frac{1}{m-1}\left(\sum_{i=1}^{m}\widehat{\beta}_i\widehat{\beta}_i' - m\overline{\beta}\,\overline{\beta}\right) - \frac{1}{m}\sum_{i=1}^{m}\widehat{\mathbf{V}}_i$$

The two-step procedure begins with the usual OLS estimate of β. With an estimate of β, we may proceed by obtaining estimates of $\widehat{\mathbf{V}}_i$ and $\widehat{\Sigma}$ (and, thus, $\widehat{\mathbf{W}}_i$), and then obtain an updated estimate of β.

Swamy (1971) further points out that the matrix $\widehat{\Sigma}$ may not be positive definite, and that since the second term is of order $1/(mT)$, it is negligible in large samples. A simple and asymptotically expedient solution is to simply drop this second term and instead use

$$\widehat{\Sigma} = \frac{1}{m-1}\left(\sum_{i=1}^{m}\widehat{\beta}_i\widehat{\beta}_i' - m\overline{\beta}\,\overline{\beta}\right)$$

As a test of the model, we may look at the difference between the OLS estimate of β, ignoring the panel structure of the data and the matrix-weighted average of the panel-specific OLS estimators. The test statistic suggested by Swamy (1971) is given by

$$\chi^2_{k(m-1)} = \sum_{i=1}^{m}(\widehat{\beta}_i - \overline{\beta}^*)'\widehat{\mathbf{V}}_i^{-1}(\widehat{\beta}_i - \overline{\beta}^*) \quad \text{where} \quad \overline{\beta}^* = \left(\sum_{i=1}^{m}\widehat{\mathbf{V}}_i^{-1}\right)^{-1}\sum_{i=1}^{m}\widehat{\mathbf{V}}_i^{-1}\widehat{\beta}_i$$

Johnston (1984) has shown that the test is algebraically equivalent to testing

$$H_0 : \beta_1 = \beta_2 = \cdots = \beta_m$$

in the generalized (groupwise heteroskedastic) xtgls model, where $\mathbf{V}$ is block diagonal with ith diagonal element Π_i.

References

Greene, W. H. 2003. *Econometric Analysis.* 5th ed. Upper Saddle River, NJ: Prentice–Hall.

Grunfeld, Y. and Z. Griliches. 1960. Is aggregation necessarily bad? *Review of Economics and Statistics* 42: 1–13.

Hardin, J. W. 1996. sg62: Hildreth–Houck random coefficients model. *Stata Technical Bulletin* 33: 21–23. Reprinted in *Stata Technical Bulletin Reprints*, vol. 6, pp. 158–162.

Hildreth, C. and C. Houck. 1968. Some estimators for a linear model with random coefficients. *Journal of the American Statistical Association* 63: 584–595.

Johnston, J. 1984. *Econometric Methods*. New York: McGraw–Hill.

Swamy, P. 1970. Efficient inference in a random coefficient regression model. *Econometrica* 38: 311–323.

——. 1971. *Statistical Inference in Random Coefficient Regression Models*. New York: Springer.

Also See

Complementary: [XT] **xtdata**, [XT] **xtdes**, [XT] **xtsum**, [XT] **xttab**,
[R] **adjust**, [R] **lincom**, [R] **mfx**, [R] **nlcom**, [R] **predict**,
[R] **predictnl**, [R] **test**, [R] **testnl**, [R] **vce**

Related: [XT] **xtgee**, [XT] **xtgls**, [XT] **xtpcse**, [XT] **xtreg**, [XT] **xtregar**

Background: [U] **16.5 Accessing coefficients and standard errors**,
[U] **23 Estimation and post-estimation commands**,
[XT] **xt**

Title

> **xtreg** — Fixed-, between-, and random-effects, and population-averaged linear models

Syntax

GLS Random-effects model

> xtreg *depvar* [*varlist*] [if *exp*] [, re i(*varname*) sa <u>the</u>ta <u>level</u>(#)]
>
> xttest0

Between-effects model

> xtreg *depvar* [*varlist*] [if *exp*] , be [i(*varname*) <u>w</u>ls <u>level</u>(#)]

Fixed-effects model

> xtreg *depvar* [*varlist*] [if *exp*] , fe [i(*varname*) <u>level</u>(#)]

ML Random-effects model

> xtreg *depvar* [*varlist*] [*weight*] [if *exp*] , mle [i(*varname*) <u>nocon</u>stant <u>level</u>(#)]

Population-averaged model

> xtreg *depvar* [*varlist*] [*weight*] [if *exp*] , pa [i(*varname*) <u>nocon</u>stant <u>level</u>(#)
>
>   <u>off</u>set(*varname*) *xtgee_options*]

by ... : may be used with xtreg; see [R] **by**.

iweights, fweights, and pweights are allowed for the population-averaged model and iweights are allowed for the maximum-likelihood (ML) random-effects model; see [U] **14.1.6 weight**. Note that weights must be constant within panels.

xtreg shares the features of all estimation commands; see [U] **23 Estimation and post-estimation commands**.

Syntax for predict

For all but the population-averaged model

> predict [*type*] *newvarname* [if *exp*] [in *range*] [, *statistic* <u>nooff</u>set]

where *statistic* is

xb	$\mathbf{x}_j\mathbf{b}$, fitted values (the default)
stdp	standard error of the fitted values
ue	$u_i + e_{it}$, the combined residual
* xbu	$\mathbf{x}_j\mathbf{b} + u_i$, prediction including effect
* u	u_i, the fixed or random error component
* e	e_{it}, the overall error component

Unstarred statistics are available both in and out of sample; type predict ... if e(sample) ... if wanted only for the estimation sample. Starred statistics are calculated only for the estimation sample even when if e(sample) is not specified.

Population-averaged model

> predict [*type*] *newvarname* [if *exp*] [in *range*] [, [mu | rate | xb | stdp]
>
> <u>nooff</u>set]

These statistics are available both in and out of sample; type predict ... if e(sample) ... if wanted only for the estimation sample.

Description

xtreg fits cross-sectional time-series regression models. In particular, xtreg with the be option fits random-effects models using the between regression estimator; with the fe option, fixed-effects models (using the within regression estimator); and with the re option, random-effects models using the GLS estimator (producing a matrix-weighted average of the between and within results). Also see [XT] **xtdata** for a faster way to fit fixed- and random-effects models.

xttest0, for use after xtreg, re, presents the Breusch and Pagan (1980) Lagrange multiplier test for random effects, a test that $\text{Var}(\nu_i) = 0$.

Options

re, the default, requests the GLS random-effects estimator.

be requests the between regression estimator.

fe requests the fixed-effects (within) regression estimator.

mle requests the maximum-likelihood random-effects estimator. Note that the utility command xttest0 may not be used after xtreg, mle.

pa requests the population-averaged estimator. For linear regression, this is the same as a random-effects estimator (both interpretations hold). Note that the utility command xttest0 may not be used after xtreg, pa.

xtreg, pa is equivalent to xtgee, family(gaussian) link(id) corr(exchangeable), which are the defaults for the xtgee command. xtreg, pa allows all the relevant xtgee options such as robust. Whether you use xtreg, pa or xtgee makes no difference. See [XT] **xtgee**.

i(*varname*) specifies the variable name that contains the unit to which the observation belongs. You can specify the i() option the first time you estimate, or you can use the iis command to set i() beforehand. Note that it is not necessary to specify i() if the data have been previously tsset, or if iis has been previously specified—in these cases, the group variable is taken from the previous setting. See [XT] **xt**.

sa specifies that the small-sample Swamy–Arora estimator individual-level variance component is to be used instead of the default consistent estimator. See the *Methods and Formulas* section for details.

theta, used with xtreg, re only, specifies that the output should include the estimated value of θ used in combining the between and fixed estimators. For balanced data, this is a constant, and for unbalanced data, a summary of the values is presented in the header of the output.

wls, used with xtreg, be only, specifies that, in the case of unbalanced data, weighted least squares be used rather than the default OLS. Both methods produce consistent estimates. The true variance of the between-effects residual is $\sigma_\nu^2 + T_i\sigma_\epsilon^2$ (see *Methods and Formulas* below). WLS produces a "stabilized" variance of $\sigma_\nu^2/T_i + \sigma_\epsilon^2$, which is also not constant. Thus, the choice between OLS and WLS amounts to which is more stable.

Comment: xtreg, be is rarely used anyway, but between estimates are an ingredient in the random-effects estimate. Our implementation of xtreg, re uses the OLS estimates for this ingredient based on our judgment that σ_ν^2 is large relative to σ_ϵ^2 in most models. Formally, any consistent estimate of the between estimates is all that is required.

noconstant suppresses the constant term (intercept) in the model.

level(#) specifies the confidence level, in percent, for confidence intervals. The default is level(95) or as set by set level; see [U] **23.6 Specifying the width of confidence intervals**.

offset(*varname*) specifies that *varname* is to be included in the model with its coefficient constrained to be 1.

xtgee_options specifies any other options allowed by xtgee for family(gaussian) link(id) such as corr(); see [XT] **xtgee**.

Options for predict

xb calculates the linear prediction; that is, $a + \mathbf{b}\mathbf{x}_{it}$. This is the default for all except the population-averaged model.

mu and rate both calculate the predicted probability of *depvar*. mu takes into account the offset(), and rate ignores those adjustments. mu and rate are equivalent if you did not specify offset(). mu is the default for the population-averaged model.

stdp calculates the standard error of the linear prediction. Note that in the case of the fixed-effects model, this excludes the variance due to uncertainty about the estimate of u_i.

ue calculates the prediction of $u_i + e_{it}$.

xbu calculates the prediction of $a + \mathbf{b}\mathbf{x}_{it} + u_i$, the prediction including the fixed or random component.

u calculates the prediction of u_i, the estimated fixed or random effect.

e calculates the prediction of e_{it}.

nooffset is relevant only if you specified offset(*varname*) for xtreg, pa. It modifies the calculations made by predict so that they ignore the offset variable; the linear prediction is treated as $\mathbf{x}_{it}\mathbf{b}$ rather than $\mathbf{x}_{it}\mathbf{b} + \text{offset}_{it}$.

Remarks

If you have not read [XT] **xt**, please do so.

See Wooldridge (2002, Chapter 10) for a good overview of fixed-effects and random-effects models.

Consider fitting models of the form

$$y_{it} = \alpha + \mathbf{x}_{it}\boldsymbol{\beta} + \nu_i + \epsilon_{it} \tag{1}$$

In this model, $\nu_i + \epsilon_{it}$ is the residual in the sense that we have little interest in it; we want estimates of $\boldsymbol{\beta}$. ν_i is the unit-specific residual; it differs between units, but, for any particular unit, its value is constant. In the pulmonary data of [XT] **xt**, a person who exercises less would presumably have a lower FEV year after year, and so would have a negative ν_i.

ϵ_{it} is the "usual" residual with the usual properties (mean 0, uncorrelated with itself, uncorrelated with $\mathbf{x}$, uncorrelated with ν, and homoskedastic), although in a more thorough development, we could decompose $\epsilon_{it} = \upsilon_t + \omega_{it}$, assume ω_{it} is a standard residual, and better describe υ_t.

Before making the assumptions necessary for estimation, let us perform some useful algebra on (1). Whatever the properties of ν_i and ϵ_{it}, if (1) is true, it must also be true that

$$\overline{y}_i = \alpha + \overline{\mathbf{x}}_i \boldsymbol{\beta} + \nu_i + \overline{\epsilon}_i \tag{2}$$

where $\overline{y}_i = \sum_t y_{it}/T_i$, $\overline{\mathbf{x}}_i = \sum_t \mathbf{x}_{it}/T_i$, and $\overline{\epsilon}_i = \sum_t \epsilon_{it}/T_i$. Subtracting (2) from (1), it must be equally true that

$$(y_{it} - \overline{y}_i) = (\mathbf{x}_{it} - \overline{\mathbf{x}}_i)\boldsymbol{\beta} + (\epsilon_{it} - \overline{\epsilon}_i) \tag{3}$$

These three equations provide the basis for estimating $\boldsymbol{\beta}$. In particular, `xtreg, fe` provides what is known as the fixed-effects estimator—also known as the within estimator—and amounts to using OLS to perform the estimation of (3). `xtreg, be` provides what is known as the between estimator, and amounts to using OLS to perform the estimation of (2). `xtreg, re` provides the random-effects estimator and is a (matrix) weighted average of the estimates produced by the between and within estimators. In particular, the random-effects estimator turns out to be equivalent to estimation of

$$(y_{it} - \theta\overline{y}_i) = (1-\theta)\alpha + (\mathbf{x}_{it} - \theta\overline{\mathbf{x}}_i)\boldsymbol{\beta} + \{(1-\theta)\nu_i + (\epsilon_{it} - \theta\overline{\epsilon}_i)\} \tag{4}$$

where θ is a function of σ_ν^2 and σ_ϵ^2. If $\sigma_\nu^2 = 0$, meaning ν_i is always 0, $\theta = 0$ and (1) can be estimated by OLS directly. Alternatively, if $\sigma_\epsilon^2 = 0$, meaning ϵ_{it} is 0, $\theta = 1$ and the within estimator returns all the information available (which will, in fact, be an $R^2 = 1$ regression).

Returning to more reasonable cases, few assumptions are required to justify the fixed-effects estimator of (3). The estimates are, however, conditional on the sample in that ν_i are not assumed to have a distribution, but are instead treated as fixed and estimable. This statistical fine point can lead to difficulty when making out-of-sample predictions, but, that aside, the fixed-effects estimator has much to recommend it.

More is required to justify the between estimator of (2), but the conditioning on the sample is not assumed since $\nu_i + \overline{\epsilon}_i$ is treated as a residual. Newly required is that we assume ν_i and $\overline{\mathbf{x}}_i$ are uncorrelated. This follows from the assumptions of the OLS estimator but is also transparent: Were ν_i and $\overline{\mathbf{x}}_i$ correlated, the estimator could not determine how much of the change in $\overline{y}_i$, associated with an increase in $\overline{\mathbf{x}}_i$, to assign to $\boldsymbol{\beta}$ versus how much to attribute to the unknown correlation. (This, of course, suggests the use of an instrumental-variable estimator, $\overline{\mathbf{z}}_i$, which is correlated with $\overline{\mathbf{x}}_i$ but uncorrelated with ν_i, but that approach is not implemented here.)

The random-effects estimator of (4) requires the same no-correlation assumption. In comparison with the between estimator, the random-effects estimator produces more efficient results, albeit ones with unknown small-sample properties. The between estimator is less efficient because it discards the over-time information in the data in favor of simple means; the random-effects estimator uses both the within and the between information.

All of this would seem to leave the between estimator of (2) with no role (except for a minor, technical part it plays in helping to estimate σ_ν^2 and σ_ϵ^2, which are used in the calculation of θ, on which the random-effects estimates depend). Let us, however, consider a variation on (1):

$$y_{it} = \alpha + \overline{\mathbf{x}}_i \boldsymbol{\beta}_1 + (\mathbf{x}_{it} - \overline{\mathbf{x}}_i)\boldsymbol{\beta}_2 + \nu_i + \epsilon_{it} \tag{1'}$$

In this model, we postulate that changes in the average value of $\mathbf{x}$ for an individual have a different effect from temporary departures from the average. In an economic situation, y might be purchases of some item and $\mathbf{x}$ income; a change in average income should have more effect than a transitory change. In a clinical situation, y might be a physical response and $\mathbf{x}$ the level of a chemical in the brain; the model allows a different response to permanent rather than transitory changes.

The variations of (2) and (3) corresponding to (1′) are

$$\overline{y}_i = \alpha + \overline{\mathbf{x}}_i\boldsymbol{\beta}_1 + \nu_i + \overline{\epsilon}_i \tag{2′}$$

$$(y_{it} - \overline{y}_i) = (\mathbf{x}_{it} - \overline{\mathbf{x}}_i)\boldsymbol{\beta}_2 + (\epsilon_{it} - \overline{\epsilon}_i) \tag{3′}$$

That is, the between estimator estimates $\boldsymbol{\beta}_1$ and the within $\boldsymbol{\beta}_2$, and neither estimates the other. Thus, even when estimating equations like (1), it is worth comparing the within and between estimators. Differences in results can suggest models like (1′), or, at the least, some other specification error.

Finally, it is worth understanding the role of the between and within estimators with regressors that are constant over time or constant over units. Consider the model

$$y_{it} = \alpha + \mathbf{x}_{it}\boldsymbol{\beta}_1 + \mathbf{s}_i\boldsymbol{\beta}_2 + \mathbf{z}_t\boldsymbol{\beta}_3 + \nu_i + \epsilon_{it} \tag{1″}$$

This model is the same as (1), except that we explicitly identify the variables that vary over both time and i ($\mathbf{x}_{it}$, such as output or FEV); variables that are constant over time ($\mathbf{s}_i$, such as race or sex); and variables that vary solely over time ($\mathbf{z}_t$, such as the consumer price index or age in a cohort study). The corresponding between and within equations are

$$\overline{y}_i = \alpha + \overline{\mathbf{x}}_i\boldsymbol{\beta}_1 + \mathbf{s}_i\boldsymbol{\beta}_2 + \overline{\mathbf{z}}\boldsymbol{\beta}_3 + \nu_i + \overline{\epsilon}_i \tag{2″}$$

$$(y_{it} - \overline{y}_i) = (\mathbf{x}_{it} - \overline{\mathbf{x}}_i)\boldsymbol{\beta}_1 + (\mathbf{z}_t - \overline{\mathbf{z}})\boldsymbol{\beta}_3 + (\epsilon_{it} - \overline{\epsilon}_i) \tag{3″}$$

In the between estimator of (2″), no estimate of $\boldsymbol{\beta}_3$ is possible because $\overline{\mathbf{z}}$ is a constant across the i observations; the regression-estimated intercept will be an estimate of $\alpha + \overline{\mathbf{z}}\boldsymbol{\beta}_3$. On the other hand, it is able to provide estimates of $\boldsymbol{\beta}_1$ and $\boldsymbol{\beta}_2$. It is able to estimate effects of factors that are constant over time, such as race and sex, but to do so, it must assume that ν_i is uncorrelated with those factors.

The within estimator of (3″), like the between estimator, provides an estimate of $\boldsymbol{\beta}_1$, but provides no estimate of $\boldsymbol{\beta}_2$ for time-invariant factors. Instead, it provides an estimate of $\boldsymbol{\beta}_3$, the effects of the time-varying factors. The between estimator can also provide estimates u_i for ν_i. More correctly, the estimator u_i is an estimator of $\nu_i + \mathbf{s}_i\boldsymbol{\beta}_2$. Thus, u_i is an estimator of ν_i only if there are no time-invariant variables in the model. If there are time-invariant variables, u_i is an estimate of ν_i plus the effects of the time-invariant variables.

Assessing goodness of fit

R^2 is a popular measure of goodness of fit in ordinary regression. In our case, given $\widehat{\alpha}$ and $\widehat{\boldsymbol{\beta}}$ estimates of α and β, we can assess the goodness of fit with respect to (1), (2), or (3). The prediction equations are, respectively,

$$\widehat{y}_{it} = \widehat{\alpha} + \mathbf{x}_{it}\widehat{\boldsymbol{\beta}} \tag{1‴}$$

$$\widehat{\overline{y}}_i = \widehat{\alpha} + \overline{\mathbf{x}}_i\widehat{\boldsymbol{\beta}} \tag{2‴}$$

$$\widehat{\widetilde{y}}_{it} = (\widehat{y}_{it} - \widehat{\overline{y}}_i) = (\mathbf{x}_{it} - \overline{\mathbf{x}}_i)\widehat{\boldsymbol{\beta}} \tag{3‴}$$

xtreg reports "R-squareds" corresponding to these three equations. R-squareds is in quotes because the R-squareds reported do not have all the properties of the OLS R^2.

The ordinary properties of R^2 include being equal to the squared correlation between $\widehat{y}$ and y and being equal to the fraction of the variation in y explained by $\widehat{y}$—formally defined as $\mathrm{Var}(\widehat{y})/\mathrm{Var}(y)$. The identity of the definitions is due to a special property of the OLS estimates; in general, given a prediction $\widehat{y}$ for y, the squared correlation is not equal to the ratio of the variances, and the ratio of the variances is not required to be less than 1.

xtreg reports R^2 values calculated as correlations squared, calling them R^2 overall, corresponding to $(1''')$; R^2 between, corresponding to $(2''')$; and R^2 within, corresponding to $(3''')$. In fact, you can think of each of these three numbers as having all the properties of ordinary R^2s if you bear in mind that the prediction being judged is not $\widehat{y}_{it}$, $\overline{\widehat{y}}_i$, and $\widehat{\widetilde{y}}_{it}$, but $\gamma_1 \widehat{y}_{it}$ from the regression $y_{it} = \gamma_1 \widehat{y}_{it}$; $\gamma_2 \overline{\widehat{y}}_i$ from the regression $\overline{y}_i = \gamma_2 \overline{\widehat{y}}_i$; and $\gamma_3 \widehat{\widetilde{y}}_{it}$ from $\widetilde{y}_{it} = \gamma_3 \widehat{\widetilde{y}}_{it}$.

In particular, xtreg, be obtains its estimates by performing OLS on (2), and therefore its reported R^2 between is an ordinary R^2. The other two reported R^2s are merely correlations squared, or, if you prefer, R^2s from the second-round regressions $y_{it} = \gamma_{11} \widehat{y}_{it}$ and $\widetilde{y}_{it} = \gamma_{13} \widehat{\widetilde{y}}_{it}$.

xtreg, fe obtains its estimates by performing OLS on (3), so its reported R^2 within is an ordinary R^2. As with be, the other R^2s are correlations squared, or, if you prefer, R^2s from the second-round regressions $\overline{y}_i = \gamma_{22} \overline{\widehat{y}}_i$ and, as with be, $\widetilde{y}_{it} = \gamma_{23} \widehat{\widetilde{y}}_{it}$.

xtreg, re obtains its estimates by performing OLS on (4); none of the R^2s corresponding to $(1''')$, $(2''')$, or $(3''')$ correspond directly to this estimator (the "relevant" R^2 is the one corresponding to (4)). All three reported R^2s are correlations squared, or, if you prefer, from second-round regressions.

xtreg and associated commands

▷ Example

Using the nlswork dataset described in [XT] **xt**, we will model ln_wage in terms of completed years of schooling (grade), current age and age squared, current years worked (experience) and experience squared, current years of tenure on the current job and tenure squared, whether black, whether resides in an area not designated an SMSA (standard metropolitan statistical area), and whether resides in the South. Most of these variables are in the data, but we need to construct a few:

```
. use http://www.stata-press.com/data/r8/nlswork
(National Longitudinal Survey.  Young Women 14-26 years of age in 1968)
. generate age2 = age^2
(24 missing values generated)
. generate ttl_exp2 = ttl_exp^2
. generate tenure2 = tenure^2
(433 missing values generated)
. generate byte black = race==2
```

(Continued on next page)

To obtain the between-effects estimates, we use xtreg, be:

```
. xtreg ln_w grade age* ttl_exp* tenure* black not_smsa south, be i(idcode)
```

Between regression (regression on group means)	Number of obs	=	28091
Group variable (i): idcode	Number of groups	=	4697

R-sq:	within = 0.1591	Obs per group: min =	1
	between = 0.4900	avg =	6.0
	overall = 0.3695	max =	15

	F(10,4686)	=	450.23
sd(u_i + avg(e_i.))= .3036114	Prob > F	=	0.0000

ln_wage	Coef.	Std. Err.	t	P>|t|	[95% Conf. Interval]	
grade	.0607602	.0020006	30.37	0.000	.0568382	.0646822
age	.0323158	.0087251	3.70	0.000	.0152105	.0494211
age2	-.0005997	.0001429	-4.20	0.000	-.0008799	-.0003194
ttl_exp	.0138853	.0056749	2.45	0.014	.0027598	.0250108
ttl_exp2	.0007342	.0003267	2.25	0.025	.0000936	.0013747
tenure	.0698419	.0060729	11.50	0.000	.0579361	.0817476
tenure2	-.0028756	.0004098	-7.02	0.000	-.0036789	-.0020722
black	-.0564167	.0105131	-5.37	0.000	-.0770272	-.0358061
not_smsa	-.1860406	.0112495	-16.54	0.000	-.2080949	-.1639862
south	-.0993378	.010136	-9.80	0.000	-.1192091	-.0794665
_cons	.3339113	.1210434	2.76	0.006	.0966093	.5712133

The between-effects regression is estimated on person averages, so it is the "n = 4697" that is relevant. xtreg, be reports the "number of observations" and group-size information; to wit: describe in [XT] **xt** showed that we have 28,534 "observations"—person-years, really—of data. Taking the subsample that has no missing values in ln_wage, grade, ..., south leaves us with 28,091 observations on person-years, reflecting 4,697 persons each observed for an average of 5.98 years.

In terms of goodness of fit, it is the R^2 between that is directly relevant; our R^2 is .4900. If, however, we use these estimates to predict the within model, we have an R^2 of .1591. If we use these estimates to fit the overall data, our R^2 is .3695.

The F statistic is a test that the coefficients on the regressors grade, age, ..., south are all jointly zero. Our model is significant.

The root mean square error of the fitted regression, which is an estimate of the standard deviation of $\nu_i + \bar{\epsilon}_i$, is .3036.

In terms of our coefficients, we find that each year of schooling increases hourly wages by 6.1%; that age increases wages up to age 26.9 and thereafter decreases them (because quadratic $ax^2 + bx + c$ turns over at $x = -b/2a$, which for our age and age2 coefficients is $.0323158/(2 \times .0005997) \approx 26.9$); that total experience increases wages at an increasing rate (which is surprising and bothersome); that tenure on the current job increases wages up to a tenure of 12.1 years and thereafter decreases them; that wages of blacks are, these things held constant, (approximately) 5.6% below that of nonblacks (approximately because black is an indicator variable); that residing in a nonSMSA (rural area) reduces wages by 18.6%; and that residing in the South reduces wages by 9.9%.

◁

▷ Example

To fit the same model with the fixed-effects estimator, we specify the fe option.

```
. xtreg ln_w grade age* ttl_exp* tenure* black not_smsa south, fe

Fixed-effects (within) regression          Number of obs      =      28091
Group variable (i): idcode                 Number of groups   =       4697

R-sq:  within  = 0.1727                     Obs per group: min =          1
       between = 0.3505                                    avg =        6.0
       overall = 0.2625                                    max =         15

                                            F(8,23386)         =     610.12
corr(u_i, Xb)  = 0.1936                      Prob > F           =     0.0000
```

ln_wage	Coef.	Std. Err.	t	P>|t|	[95% Conf. Interval]	
grade	(dropped)					
age	.0359987	.0033864	10.63	0.000	.0293611	.0426362
age2	-.000723	.0000533	-13.58	0.000	-.0008274	-.0006186
ttl_exp	.0334668	.0029653	11.29	0.000	.0276545	.039279
ttl_exp2	.0002163	.0001277	1.69	0.090	-.0000341	.0004666
tenure	.0357539	.0018487	19.34	0.000	.0321303	.0393775
tenure2	-.0019701	.000125	-15.76	0.000	-.0022151	-.0017251
black	(dropped)					
not_smsa	-.0890108	.0095316	-9.34	0.000	-.1076933	-.0703282
south	-.0606309	.0109319	-5.55	0.000	-.0820582	-.0392036
_cons	1.03732	.0485546	21.36	0.000	.9421497	1.13249
sigma_u	.35562203					
sigma_e	.29068923					
rho	.59946283	(fraction of variance due to u_i)				

```
F test that all u_i=0:     F(4696,23386) =      5.13          Prob > F = 0.0000
```

The observation summary at the top is the same as for the between-effects model, although this time it is the "Number of obs" that is relevant.

Our three R^2s are not too different from those reported previously; the R^2 within is slightly higher (.1727 vs .1591) and the R^2 between a little lower (.3505 vs .4900), which is as expected since the between estimator maximizes R^2 between and the within estimator R^2 within. In terms of overall fit, these estimates are somewhat worse (.2625 vs .3695).

xtreg, fe is able to provide estimates of σ_ν and σ_ϵ, although how you interpret these estimates depends on whether you are using xtreg to fit a fixed-effects model or random-effects model. To clarify this fine point, in the fixed-effects model, ν_i are formally fixed—they have no distribution. If you subscribe to this view, think of the reported $\hat{\sigma}_\nu$ as merely an arithmetic way to describe the range of the estimated but fixed ν_i. If, however, you are employing the fixed-effects estimator of the random-effects model—as is likely—then .355622 is an estimate of σ_ν, or it would be if there were no dropped variables in the estimation.

In our case, note that both grade and black were dropped from the model. They were dropped because they do not vary over time. Since grade and race are time-invariant, our estimate u_i is an estimate of ν_i plus the effects of grade and race, and so our estimate of the standard deviation is based on the variation in ν_i, grade, and race. On the other hand, had race and grade been dropped merely because they were collinear with the other regressors in our model, u_i would be an estimate of ν_i and .3556 would be an estimate of σ_ν. (xtsum and xttab allow determining whether a variable is time-invariant; see [XT] **xtsum** and [XT] **xttab**.)

Regardless of the status of our estimator u_i, our estimate of the standard deviation of ϵ_{it} is valid (and, in fact, is the estimate that would be used by the random-effects estimator in producing its results).

Our estimate of the correlation of u_i with $\mathbf{x}_{it}$ suffers from the problem of what u_i measures.

We find correlation, but cannot say whether this is correlation of ν_i with $\mathbf{x}_{it}$ or merely correlation of grade and race. In any case, the fixed-effects estimator is robust to such a correlation, and the other estimates it produces are unbiased.

So, while this estimator produces no estimates of the effects of grade and race, it does predict that age has a positive effect on wages up to age 24.9 years (as compared with 26.9 years estimated by the between estimator); that total experience still increases wages at an increasing rate (which is still bothersome); that tenure increases wages up to 9.1 years (as compared with 12.1); that living in a nonSMSA reduces wages by 8.9% (as compared with a more drastic 18.6%); and that living in the South reduces wages by 6.1% (as compared with 9.9%).

◁

▷ Example

Refitting our log-wage model with the random-effects estimator, we obtain

```
. xtreg ln_w grade age* ttl_exp* tenure* black not_smsa south, re
```

| Random-effects GLS regression | | | | Number of obs | = | 28091 |
| Group variable (i): idcode | | | | Number of groups | = | 4697 |

R-sq: within = 0.1715				Obs per group: min =		1
between = 0.4784				avg =		6.0
overall = 0.3708				max =		15

| Random effects u_i ~ Gaussian | | | | Wald chi2(10) | = | 9244.87 |
| corr(u_i, X)   = 0 (assumed) | | | | Prob > chi2 | = | 0.0000 |

ln_wage	Coef.	Std. Err.	z	P>|z|	[95% Conf. Interval]	
grade	.0646499	.0017811	36.30	0.000	.0611589	.0681408
age	.036806	.0031195	11.80	0.000	.0306918	.0429201
age2	-.0007133	.00005	-14.27	0.000	-.0008113	-.0006153
ttl_exp	.0290207	.0024219	11.98	0.000	.0242737	.0337676
ttl_exp2	.0003049	.0001162	2.62	0.009	.000077	.0005327
tenure	.039252	.0017555	22.36	0.000	.0358114	.0426927
tenure2	-.0020035	.0001193	-16.80	0.000	-.0022373	-.0017697
black	-.0530532	.0099924	-5.31	0.000	-.0726379	-.0334685
not_smsa	-.1308263	.0071751	-18.23	0.000	-.1448891	-.1167634
south	-.0868927	.0073031	-11.90	0.000	-.1012066	-.0725788
_cons	.2387209	.0494688	4.83	0.000	.1417639	.335678

sigma_u	.25790313					
sigma_e	.29069544					
rho	.44043812	(fraction of variance due to u_i)				

According to the R^2s, this estimator performs worse within than the within/fixed-effects estimator and worse between than the between estimator, as it must, and slightly better overall.

We estimate that σ_ν is .2579 and σ_ϵ is .2907, and, by assertion, assume that the correlation of ν and $\mathbf{x}$ is zero.

All that is known about the random-effects estimator is its asymptotic properties, so rather than reporting an F statistic for overall significance, xtreg, re reports a χ^2. Taken jointly, our coefficients are significant.

Also reported is a summary of the distribution of θ_i, an ingredient in the estimation of (4). θ is not a constant in this case because we observe women for unequal periods of time.

In terms of interpretation, we estimate that schooling has a rate of return of 6.5% (compared with 6.1% between and no estimate within); that the increase of wages with age turns around at 25.8 years (compared with 26.9 between and 24.9 within); that total experience yet again increases wages increasingly; that the effect of job tenure turns around at 9.8 years (compared with 12.1 between and 9.1 within); that being black reduces wages by 5.3% (compared with 5.6% between and no estimate within); that living in a nonSMSA reduces wages 13.1% (compared with 18.6% between and 8.9% within); and that living in the South reduces wages 8.7% (compared with 9.9% between and 6.1% within).

◁

▷ Example

Alternatively, we could have fitted this random-effects model using the maximum likelihood estimator:

```
. xtreg ln_w grade age* ttl_exp* tenure* black not_smsa south, mle
Fitting constant-only model:
Iteration 0:    log likelihood = -13690.161
Iteration 1:    log likelihood = -12819.317
Iteration 2:    log likelihood = -12662.039
Iteration 3:    log likelihood = -12649.744
Iteration 4:    log likelihood = -12649.614

Fitting full model:
Iteration 0:    log likelihood =  -8922.145
Iteration 1:    log likelihood = -8853.6409
Iteration 2:    log likelihood = -8853.4255
Iteration 3:    log likelihood = -8853.4254
```

Random-effects ML regression	Number of obs	=	28091
Group variable (i): idcode	Number of groups	=	4697
Random effects u_i ~ Gaussian	Obs per group: min =		1
	avg =		6.0
	max =		15
	LR chi2(10)	=	7592.38
Log likelihood = -8853.4254	Prob > chi2	=	0.0000

ln_wage	Coef.	Std. Err.	z	P>|z|	[95% Conf. Interval]	
grade	.0646093	.0017372	37.19	0.000	.0612044	.0680142
age	.0368531	.0031226	11.80	0.000	.030733	.0429732
age2	-.0007132	.0000501	-14.24	0.000	-.0008113	-.000615
ttl_exp	.0288196	.0024143	11.94	0.000	.0240877	.0335515
ttl_exp2	.000309	.0001163	2.66	0.008	.0000811	.0005369
tenure	.0394371	.0017604	22.40	0.000	.0359868	.0428875
tenure2	-.0020052	.0001195	-16.77	0.000	-.0022395	-.0017709
black	-.0533394	.0097338	-5.48	0.000	-.0724172	-.0342615
not_smsa	-.1323433	.0071322	-18.56	0.000	-.1463221	-.1183644
south	-.0875599	.0072143	-12.14	0.000	-.1016998	-.0734201
_cons	.2390837	.0491902	4.86	0.000	.1426727	.3354947
/sigma_u	.2485556	.0035017	70.98	0.000	.2416925	.2554187
/sigma_e	.2918458	.001352	215.87	0.000	.289196	.2944956
rho	.4204033	.0074828			.4057959	.4351212

Likelihood-ratio test of sigma_u=0: chibar2(01)= 7339.84 Prob>=chibar2 = 0.000

The estimates are very nearly the same as those produced by `xtreg, re`—the GLS estimator. For instance, `xtreg, re` estimated the coefficient on `grade` to be .0646499; `xtreg, mle` estimated .0646093; and the ratio is .0646499/.0646093 = 1.001 to three decimal places. Similarly, the standard errors are nearly equal: .0017812/.0017372 = 1.025. Below we compare all 11 coefficients:

Estimator	Coefficient ratio			SE ratio		
	mean	min.	max.	mean	min.	max.
`xtreg, mle` (ML)	1.	1.	1.	1.	1.	1.
`xtreg, re` (GLS)	.997	.987	1.007	1.006	.997	1.027

◁

▷ Example

We could also have fitted this model using the population-averaged estimator:

```
. xtreg ln_w grade age* ttl_exp* tenure* black not_smsa south, i(idcode) pa

Iteration 1: tolerance = .0310561
Iteration 2: tolerance = .00074898
Iteration 3: tolerance = .0000147
Iteration 4: tolerance = 2.880e-07

GEE population-averaged model            Number of obs      =      28091
Group variable:               idcode     Number of groups   =       4697
Link:                       identity     Obs per group: min =          1
Family:                     Gaussian                    avg =        6.0
Correlation:            exchangeable                    max =         15
                                         Wald chi2(10)      =    9598.89
Scale parameter:              .1436709   Prob > chi2        =     0.0000
```

ln_wage	Coef.	Std. Err.	z	P>|z|	[95% Conf. Interval]	
grade	.0645427	.0016829	38.35	0.000	.0612442	.0678412
age	.036932	.0031509	11.72	0.000	.0307564	.0431076
age2	-.0007129	.0000506	-14.10	0.000	-.0008121	-.0006138
ttl_exp	.0284878	.0024169	11.79	0.000	.0237508	.0332248
ttl_exp2	.0003158	.0001172	2.69	0.007	.000086	.0005456
tenure	.0397468	.0017779	22.36	0.000	.0362621	.0432315
tenure2	-.002008	.0001209	-16.61	0.000	-.0022449	-.0017711
black	-.0538314	.0094086	-5.72	0.000	-.072272	-.0353909
not_smsa	-.1347788	.0070543	-19.11	0.000	-.1486049	-.1209526
south	-.0885969	.0071132	-12.46	0.000	-.1025386	-.0746552
_cons	.2396286	.0491465	4.88	0.000	.1433034	.3359539

These results differ from those produced by `xtreg, re` and `xtreg, mle`. Coefficients are larger and standard errors smaller. `xtreg, pa` is simply another way to run the `xtgee` command. That is, we would have obtained the same output had we typed

```
. xtgee ln_w grade age* ttl_exp* tenure* black not_smsa south, i(idcode)
```
(output omitted because it is the same as above)

See [XT] **xtgee**. In the language of `xtgee`, the random-effects model corresponds to an `exchangeable` correlation structure and `identity` link, and `xtgee` has the advantage that it will allow other correlation structures as well. Let us stay with the random-effects model, however. `xtgee` will also produce robust estimates of variance, and we refit this model that way by typing

. xtgee ln_w grade age* ttl_exp* tenure* black not_smsa south, i(idcode) robust
(*output omitted, coefficients the same, standard errors different*)

In the previous example, we presented a table comparing xtreg, re with xtreg, mle. Below we add the results from the estimates shown and the ones we did with xtgee, robust:

Estimator		Coefficient ratio			SE ratio		
		mean	min.	max.	mean	min.	max.
xtreg, mle	(ML)	1.	1.	1.	1.	1.	1.
xtreg, re	(GLS)	.997	.987	1.007	1.006	.997	1.027
xtreg, pa	(PA)	1.060	.847	1.317	.853	.626	.986
xtgee, robust	(PA)	1.060	.847	1.317	1.306	.957	1.545

So which are right? This is a real dataset and we do not know. However, in the example after the next, we will present evidence that the assumptions underlying the xtreg, re and xtreg, mle results are not met.

◁

▷ Example

After xtreg, re estimation, xttest0 will report a test of $\nu_i = 0$, in case we have any doubts:

. xttest0

Breusch and Pagan Lagrangian multiplier test for random effects:

        ln_wage[idcode,t] = Xb + u[idcode] + e[idcode,t]

        Estimated results:

	Var	sd = sqrt(Var)
ln_wage	.2283326	.4778416
e	.0845038	.2906954
u	.066514	.2579031

    Test: Var(u) = 0
                        chi2(1) = 14779.98
                    Prob > chi2 =   0.0000

◁

▷ Example

More importantly, after xtreg, re estimation, hausman will perform the Hausman specification test. If our model is correctly specified and if ν_i is uncorrelated with $\mathbf{x}_{it}$, then the (subset of) coefficients that are estimated by the fixed-effects estimator and the same coefficients that are estimated here should not statistically differ:

(*Continued on next page*)

```
. xtreg ln_w grade age* ttl_exp* tenure* black not_smsa south, re
(output omitted)

. estimates store random_effects

. xtreg ln_w grade age* ttl_exp* tenure* black not_smsa south, fe
(output omitted)

. hausman . random_effects
```

| | ―― Coefficients ―― | | | |
| | (b) | (B) | (b-B) | sqrt(diag(V_b-V_B)) |
	.	random_eff~s	Difference	S.E.
age	.0359987	.036806	-.0008073	.0013177
age2	-.000723	-.0007133	-9.68e-06	.0000184
ttl_exp	.0334668	.0290207	.0044461	.001711
ttl_exp2	.0002163	.0003049	-.0000886	.000053
tenure	.0357539	.039252	-.0034981	.0005797
tenure2	-.0019701	-.0020035	.0000334	.0000373
not_smsa	-.0890108	-.1308263	.0418155	.0062745
south	-.0606309	-.0868927	.0262618	.0081346

```
                         b = consistent under Ho and Ha; obtained from xtreg
            B = inconsistent under Ha, efficient under Ho; obtained from xtreg
    Test:  Ho:  difference in coefficients not systematic
                chi2(8) = (b-B)'[(V_b-V_B)^(-1)](b-B)
                        =       149.44
                Prob>chi2 =       0.0000
```

We can reject the hypothesis that the coefficients are the same. Before turning to what this means, note that hausman listed the coefficients estimated by the two models. It did not, however, list grade and race. hausman did not make a mistake; in the Hausman test, one compares only the coefficients estimated by both techniques.

What does this mean? We have an unpleasant choice: we can admit that our model is misspecified—that we have not parameterized it correctly—or we can hold to our specification being correct, in which case the observed differences must be due to the zero-correlation of ν_i and the $\mathbf{x}_{it}$ assumption.

◁

❑ Technical Note

We can also mechanically explore the underpinnings of the test's dissatisfaction. In the comparison table from hausman, note that it is the coefficients on not_smsa and south that exhibit the largest differences. In $(1')$, we showed how to decompose a model into within and between effects. Let us do that with these two variables, assuming that changes in the average have one effect while transitional changes have another:

```
. egen avgnsmsa = mean(not_smsa), by(idcode)

. generate devnsma = not_smsa -avgnsmsa
(8 missing values generated)

. egen avgsouth = mean(south), by(idcode)

. generate devsouth = south - avgsouth
(8 missing values generated)
```

```
. xtreg ln_w grade age* ttl_exp* tenure* black avgnsm devnsm avgsou devsou
```

Random-effects GLS regression	Number of obs	=	28091
Group variable (i): idcode	Number of groups	=	4697

R-sq: within = 0.1723	Obs per group: min =	1
between = 0.4809	avg =	6.0
overall = 0.3737	max =	15

Random effects u_i ~ Gaussian	Wald chi2(12)	=	9319.69
corr(u_i, X) = 0 (assumed)	Prob > chi2	=	0.0000

ln_wage	Coef.	Std. Err.	z	P>|z|	[95% Conf. Interval]	
grade	.0631716	.0017903	35.29	0.000	.0596627	.0666805
age	.0375196	.0031186	12.03	0.000	.0314072	.043632
age2	-.0007248	.00005	-14.50	0.000	-.0008228	-.0006269
ttl_exp	.0286542	.0024207	11.84	0.000	.0239097	.0333987
ttl_exp2	.0003222	.0001162	2.77	0.006	.0000945	.0005499
tenure	.0394424	.001754	22.49	0.000	.0360045	.0428803
tenure2	-.0020081	.0001192	-16.85	0.000	-.0022417	-.0017746
black	-.0545938	.0102099	-5.35	0.000	-.0746048	-.0345827
avgnsmsa	-.1833238	.0109337	-16.77	0.000	-.2047533	-.1618942
devnsma	-.0887596	.0095071	-9.34	0.000	-.1073932	-.070126
avgsouth	-.1011235	.0098787	-10.24	0.000	-.1204855	-.0817616
devsouth	-.0598538	.0109054	-5.49	0.000	-.081228	-.0384796
_cons	.268298	.0495776	5.41	0.000	.1711277	.3654683

sigma_u	.25791607	
sigma_e	.29069544	
rho	.44046285	(fraction of variance due to u_i)

We will leave the reinterpretation of this model to you, except to note that if we were really going to sell this model, we would have to explain why the between and within effects are different. Focusing on residence in a nonSMSA, we might tell a story about rural folk being paid less and continuing to get paid less when they move to the SMSA. As such, however, it is just a story. Given our cross-sectional time-series data, we could create variables to measure this (an indicator for moved from nonSMSA to SMSA) and to measure the effects. In our assessment of this model, we should think about women in the cities moving to the country and their relative productivity in a bucolic setting.

(Continued on next page)

In any case, the Hausman test now is

```
. estimates store new_random_effects

. xtreg ln_w grade age* ttl_exp* tenure* black avgnsm devnsm avgsou devsou, fe
(output omitted)

. hausman . new_random_effects
```

	(b)	(B) new_random~s	(b-B) Difference	sqrt(diag(V_b-V_B)) S.E.
	—— Coefficients ——			
age	.0359987	.0375196	-.001521	.0013198
age2	-.000723	-.0007248	1.84e-06	.0000184
ttl_exp	.0334668	.0286542	.0048126	.0017127
ttl_exp2	.0002163	.0003222	-.0001059	.0000531
tenure	.0357539	.0394424	-.0036885	.0005839
tenure2	-.0019701	-.0020081	.000038	.0000377
devnsma	-.0890108	-.0887596	-.0002512	.0006826
devsouth	-.0606309	-.0598538	-.0007771	.0007612

```
                      b = consistent under Ho and Ha; obtained from xtreg
           B = inconsistent under Ha, efficient under Ho; obtained from xtreg
    Test:  Ho:  difference in coefficients not systematic
              chi2(8) = (b-B)'[(V_b-V_B)^(-1)](b-B)
                      =       92.52
            Prob>chi2 =      0.0000
```

We have mechanically succeeded in greatly reducing the χ^2, but not by enough. The major differences now are in the age, experience, and tenure effects. We already knew this problem existed because of the ever-increasing effect of experience. More careful parameterization work than simply including squares needs to be done.

❑

Acknowledgments

We thank Richard Goldstein, who wrote the first draft of the routine that fits random-effects regressions, and Badi Baltagi and Manuelita Ureta of Texas A&M University, who assisted us in working our way through the literature.

(Continued on next page)

Saved Results

xtreg, re saves in e():

Scalars

e(N)	number of observations	e(r2_o)	R-squared for overall model
e(N_g)	number of groups	e(r2_b)	R-squared for between model
e(df_m)	model degrees of freedom	e(sigma)	ancillary parameter (gamma, lnormal)
e(g_max)	largest group size	e(sigma_u)	panel-level standard deviation
e(g_min)	smallest group size	e(sigma_e)	standard deviation of ϵ_{it}
e(g_avg)	average group size	e(thta_min)	minimum θ
e(chi2)	χ^2	e(thta_5)	θ, 5th percentile
e(rho)	ρ	e(thta_50)	θ, 50th percentile
e(Tbar)	harmonic mean of group sizes	e(thta_95)	θ, 95th percentile
e(Tcon)	1 if T is constant	e(thta_max)	maximum θ
e(r2_w)	R-squared for within model		

Macros

e(cmd)	xtreg	e(chi2type)	Wald; type of model χ^2 test
e(depvar)	name of dependent variable	e(sa)	Swamy–Arora estimator of the variance
e(model)	re		components (sa only)
e(ivar)	variable denoting groups	e(predict)	program used to implement predict

Matrices

e(b)	coefficient vector	e(Vf)	VCE for fixed-effects model
e(theta)	θ	e(bf)	coefficient vector for fixed-effects model
e(V)	variance–covariance matrix of the estimators		

Functions

e(sample)	marks estimation sample

xtreg, be saves in e():

Scalars

e(N)	number of observations	e(ll)	log likelihood
e(N_g)	number of groups	e(ll_0)	log likelihood, constant-only model
e(mss)	model sum of squares	e(g_max)	largest group size
e(df_m)	model degrees of freedom	e(g_min)	smallest group size
e(rss)	residual sum of squares	e(g_avg)	average group size
e(df_r)	residual degrees of freedom	e(Tbar)	harmonic mean of group sizes
e(r2)	R-squared	e(Tcon)	1 if T is constant
e(r2_a)	adjusted R-squared	e(r2_w)	R-squared for within model
e(F)	F statistic	e(r2_o)	R-squared for overall model
e(rmse)	root mean square error	e(r2_b)	R-squared for between model

Macros

e(cmd)	xtreg	e(ivar)	variable denoting groups
e(depvar)	name of dependent variable	e(predict)	program used to implement predict
e(model)	be		

Matrices

e(b)	coefficient vector	e(V)	variance–covariance matrix of the estimators

Functions

e(sample)	marks estimation sample

`xtreg, fe` saves in `e()`:

Scalars

e(N)	number of observations	e(g_max)	largest group size
e(N_g)	number of groups	e(g_min)	smallest group size
e(mss)	model sum of squares	e(g_avg)	average group size
e(tss)	total sum of squares	e(rho)	ρ
e(df_m)	model degrees of freedom	e(Tbar)	harmonic mean of group sizes
e(rss)	residual sum of squares	e(Tcon)	1 if T is constant
e(df_r)	residual degrees of freedom	e(r2_w)	R-squared for within model
e(r2)	R-squared	e(r2_o)	R-squared for overall model
e(r2_a)	adjusted R-squared	e(r2_b)	R-squared for between model
e(F)	F statistic	e(sigma)	ancillary parameter (gamma, lnormal)
e(rmse)	root mean square error	e(corr)	corr(u_i, Xb)
e(ll)	log likelihood	e(sigma_u)	panel-level standard deviation
e(ll_0)	log likelihood, constant-only model	e(sigma_e)	standard deviation of ϵ_{it}
e(df_a)	degrees of freedom for absorbed effect	e(F_f)	F for $u_i=0$

Macros

e(cmd)	xtreg	e(ivar)	variable denoting groups
e(depvar)	name of dependent variable	e(predict)	program used to implement predict
e(model)	fe		

Matrices

e(b)	coefficient vector	e(V)	variance–covariance matrix of the estimators

Functions

e(sample)	marks estimation sample

`xtreg, mle` saves in `e()`:

Scalars

e(N)	number of observations	e(g_min)	smallest group size
e(N_g)	number of groups	e(g_avg)	average group size
e(df_m)	model degrees of freedom	e(chi2)	χ^2
e(ll)	log likelihood	e(chi2_c)	χ^2 for comparison test
e(ll_0)	log likelihood, constant-only model	e(rho)	ρ
e(ll_c)	log likelihood, comparison model	e(sigma_u)	panel-level standard deviation
e(g_max)	largest group size	e(sigma_e)	standard deviation of ϵ_{it}

Macros

e(cmd)	xtreg	e(chi2type)	Wald or LR; type of model χ^2 test
e(depvar)	name of dependent variable	e(chi2_ct)	Wald or LR; type of model χ^2 test corresponding to e(chi2_c)
e(title)	title in estimation output		
e(model)	ml	e(distrib)	Gaussian; the distribution of the re
e(ivar)	variable denoting groups	e(crittype)	optimization criterion
e(wtype)	weight type	e(predict)	program used to implement predict
e(wexp)	weight expression		

Matrices

e(b)	coefficient vector	e(V)	variance–covariance matrix of the estimators

Functions

e(sample)	marks estimation sample

xtreg, pa saves in e():

Scalars

e(N)	number of observations	e(deviance)	deviance
e(N_g)	number of groups	e(chi2_dev)	χ^2 test of deviance
e(df_m)	model degrees of freedom	e(dispers)	deviance dispersion
e(g_max)	largest group size	e(chi2_dis)	χ^2 test of deviance dispersion
e(g_min)	smallest group size	e(tol)	target tolerance
e(g_avg)	average group size	e(dif)	achieved tolerance
e(chi2)	χ^2	e(phi)	scale parameter
e(df_pear)	degrees of freedom for Pearson χ^2		

Macros

e(cmd)	xtgee	e(ivar)	variable denoting groups
e(cmd2)	xtreg	e(vcetype)	covariance estimation method
e(depvar)	name of dependent variable	e(chi2type)	Wald; type of model χ^2 test
e(model)	pa	e(disp)	deviance dispersion
e(family)	Gaussian	e(offset)	offset
e(link)	identity; link function	e(crittype)	optimization criterion
e(corr)	correlation structure	e(predict)	program used to implement predict
e(scale)	x2, dev, phi, or #; scale parameter		

Matrices

e(b)	coefficient vector	e(V)	variance–covariance matrix of the estimators
e(R)	estimated working correlation matrix		

Functions

e(sample)	marks estimation sample

Methods and Formulas

The model to be fitted is

$$y_{it} = \alpha + \mathbf{x}_{it}\boldsymbol{\beta} + \nu_i + \epsilon_{it}$$

for $i = 1, \ldots, n$ and, for each i, $t = 1, \ldots, T$, of which T_i periods are actually observed.

xtreg, fe

xtreg, fe produces estimates by running OLS on

$$(y_{it} - \overline{y}_i + \overline{\overline{y}}) = \alpha + (\mathbf{x}_{it} - \overline{\mathbf{x}}_i + \overline{\overline{\mathbf{x}}})\boldsymbol{\beta} + (\epsilon_{it} - \overline{\epsilon}_i + \overline{\nu}) + \overline{\overline{\epsilon}}$$

where $\overline{y}_i = \sum_{t=1}^{T_i} y_{it}/T_i$, and similarly, $\overline{\overline{y}} = \sum_i \sum_t y_{it}/(nT_i)$. The covariance matrix of the estimators is adjusted for the extra $n - 1$ estimated means, so results are the same as using OLS on (1) to estimate ν_i directly.

From the estimates $\widehat{\alpha}$ and $\widehat{\boldsymbol{\beta}}$, estimates u_i of ν_i are obtained as $u_i = \overline{y}_i - \widehat{\alpha} - \overline{\mathbf{x}}_i\widehat{\boldsymbol{\beta}}$. Reported from the calculated u_i is its standard deviation and its correlation with $\overline{\mathbf{x}}_i\widehat{\boldsymbol{\beta}}$. Reported as the standard deviation of e_{it} is the regression's estimated root mean square error, s, which is adjusted (as previously stated) for the $n - 1$ estimated means.

Reported as R^2 within is the R^2 from the mean-deviated regression.

Reported as R^2 between is $\mathrm{corr}(\overline{\mathbf{x}}_i\widehat{\boldsymbol{\beta}}, \overline{y}_i)^2$.

Reported as R^2 overall is $\text{corr}(\mathbf{x}_{it}\widehat{\boldsymbol{\beta}}, y_{it})^2$.

xtreg, be

`xtreg, be` fits the following model:

$$\overline{y}_i = \alpha + \overline{\mathbf{x}}_i\boldsymbol{\beta} + \nu_i + \overline{\epsilon}_i$$

Estimation is via OLS unless T_i is not constant and the `wls` option is specified. Otherwise, the estimation is performed via WLS. The estimation is performed by `regress` for both cases, but in the case of WLS, `[aweight=`T_i`]` is specified.

Reported as R^2 between is the R^2 from the fitted regression.

Reported as R^2 within is $\text{corr}\big\{(\mathbf{x}_{it} - \overline{\mathbf{x}}_i)\widehat{\boldsymbol{\beta}}, y_{it} - \overline{y}_i\big\}^2$.

Reported as R^2 overall is $\text{corr}(\mathbf{x}_{it}\widehat{\boldsymbol{\beta}}, y_{it})^2$.

xtreg, re

The key to the random-effects estimator is the GLS transform. Given estimates of the idiosyncratic component, $\widehat{\sigma}_e^2$, and the individual component, $\widehat{\sigma}_u^2$, the GLS transform of a variable z for the random-effects model is

$$z_{it}^* = z_{it} - \widehat{\theta}_i\overline{z}_i$$

where $\overline{z}_i = \frac{1}{T_i}\sum_t^{T_i} z_{it}$ and

$$\widehat{\theta}_i = 1 - \sqrt{\frac{\widehat{\sigma}_e^2}{T_i\widehat{\sigma}_u^2 + \widehat{\sigma}_e^2}}$$

Given an estimate of $\widehat{\theta}_i$, one transforms the dependent and independent variables, and then the coefficient estimates and the variance–covariance matrix come from an OLS regression of y_{it}^* on $\mathbf{x}_{it}^*$ and the transformed constant $1 - \widehat{\theta}_i$.

Stata has two implementations of the Swamy–Arora method for estimating the variance components. They produce exactly the same results in balanced panels, and share the same estimator of σ_e^2. However, the two methods differ in their estimator of σ_u^2 in unbalanced panels. We call the first $\widehat{\sigma}_{u\overline{T}}^2$ and the second $\widehat{\sigma}_{uSA}^2$. Both estimators are consistent; however, $\widehat{\sigma}_{uSA}^2$ has a more elaborate adjustment for small samples than $\widehat{\sigma}_{u\overline{T}}^2$. (See Baltagi (2001), Baltagi and Chang (1994), and Swamy and Arora (1972) for derivations of these methods.)

Both methods use the same function of within residuals to estimate the idiosyncratic error component σ_e. Specifically,

$$\widehat{\sigma}_e = \frac{\sum_i^n \sum_t^{T_i} e_{it}^2}{N - n - K + 1}$$

where

$$e_{it} = (y_{it} - \overline{y}_i + \overline{\overline{y}}) - \widehat{\alpha}_w - (\mathbf{x}_{it} - \overline{\mathbf{x}}_i + \overline{\overline{\mathbf{x}}})\widehat{\boldsymbol{\beta}}_w$$

and $\widehat{\alpha}_w$ and $\widehat{\beta}_w$ are the within estimates of the coefficients and $N = \sum_i^n T_i$. Here, the intuition is straightforward. After passing the within residuals through the within transform, only the idiosyncratic errors are left.

The default method for estimating σ_u^2 is

$$\widehat{\sigma}_{u\overline{T}}^2 = \max\left\{0, \frac{\frac{SSR_b}{N-K} - \widehat{\sigma}_e^2}{\overline{T}}\right\}$$

where

$$SSR_b = \sum_i^n T_i\left(\overline{y}_i - \widehat{\alpha}_b - \overline{\mathbf{x}}_i\widehat{\beta}_b\right)^2$$

$\widehat{\alpha}_b$ and $\widehat{\beta}_b$ are coefficient estimates from the between regression and $\overline{T}$ is the harmonic mean of T_i; i.e.,

$$\overline{T} = \frac{n}{\sum_i^n \frac{1}{T_i}}$$

This estimator is consistent for σ_u^2, and is computationally less expensive than the second method. Here, the intuition is that the sum of squared residuals from the between model estimate a function of both the idiosyncratic component and the individual component. Using our estimator of σ_e^2, we can remove the idiosyncratic component, leaving only the desired individual component.

The second method is the Swamy–Arora method for unbalanced panels derived by Baltagi and Chang (1994). This method is based on the same intuition, but it has a more precise small-sample adjustment. Using this method,

$$\widehat{\sigma}_{\mathrm{uSA}} = \max\left\{0, \frac{SSR_b - (n - K)\widehat{\sigma}_e^2}{N - tr}\right\}$$

where

$$tr = \mathrm{trace}\left\{(\mathbf{X}'\mathbf{P}\mathbf{X})^{-1}\mathbf{X}'\mathbf{Z}\mathbf{Z}'\mathbf{X}\right\}$$

$$\mathbf{P} = \mathrm{diag}\left\{\left(\frac{1}{T_i}\right)\boldsymbol{\iota}_{T_i}\boldsymbol{\iota}_{T_i}'\right\}$$

$$\mathbf{Z} = \mathrm{diag}\left[\boldsymbol{\iota}_{T_i}\right]$$

$\mathbf{X}$ is the $N \times K$ matrix of covariates, including the constant, and $\boldsymbol{\iota}_{T_i}$ is a $T_i \times 1$ vector of ones.

The estimated coefficients $(\widehat{\alpha}_r, \widehat{\beta}_r)$ and their covariance matrix $\mathbf{V}_r$ are reported together with the previously calculated quantities $\widehat{\sigma}_e$ and $\widehat{\sigma}_u$. The standard deviation of $\nu_i + e_{it}$ is calculated as $\sqrt{\widehat{\sigma}_e^2 + \widehat{\sigma}_u^2}$.

Reported as R^2 between is $\mathrm{corr}(\overline{\mathbf{x}}_i\widehat{\beta}, \overline{y}_i)^2$.

Reported as R^2 within is $\mathrm{corr}\{(\mathbf{x}_{it} - \overline{\mathbf{x}}_i)\widehat{\beta}, y_{it} - \overline{y}_i\}^2$.

Reported as R^2 overall is $\mathrm{corr}(\mathbf{x}_{it}\widehat{\beta}, y_{it})^2$.

xtreg, mle

The log likelihood for the ith unit is

$$l_i = -\frac{1}{2}\left(\frac{1}{\sigma_e^2}\left[\sum_{t=1}^{T_i}(y_{it} - \mathbf{x}_{it}\boldsymbol{\beta})^2 - \frac{\sigma_u^2}{T_i\sigma_u^2 + \sigma_e^2}\left\{\sum_{t=1}^{T_i}(y_{it} - \mathbf{x}_{it}\boldsymbol{\beta})\right\}^2\right]\right.$$
$$\left. + \ln\left(T_i\frac{\sigma_u^2}{\sigma_e^2} + 1\right) + T_i\ln(2\pi\sigma_e^2)\right)$$

The mle and re options yield essentially the same results, except when total $N = \sum_i T_i$ is small (200 or less) and the data are unbalanced.

xtreg, pa

See [XT] **xtgee** for details on the methods and formulas used to calculate the population-averaged model using a generalized estimating equations approach.

xttest0

xttest0 reports the Lagrange multiplier test for random effects developed by Breusch and Pagan (1980) and as modified by Baltagi and Li (1990). The model

$$y_{it} = \alpha + \mathbf{x}_{it}\boldsymbol{\beta} + \nu_{it}$$

is estimated via OLS, and then the quantity

$$\lambda_{\mathrm{LM}} = \frac{(n\overline{T})^2}{2}\left(\frac{A_1^2}{(\sum_i T_i^2) - n\overline{T}}\right)$$

is calculated, where

$$A_1 = 1 - \frac{\sum_{i=1}^{n}(\sum_{t=1}^{T_i}v_{it})^2}{\sum_i\sum_t v_{it}^2}$$

The Baltagi and Li modification allows for unbalanced data and reduces to the standard formula

$$\lambda_{\mathrm{LM}} = \frac{nT}{2(T-1)}\left\{\frac{\sum_i(\sum_t v_{it})^2}{\sum_i\sum_t v_{it}^2} - 1\right\}^2$$

when $T_i = T$ (balanced data). Under the null hypothesis, λ_{LM} is distributed $\chi^2(1)$.

References

Baltagi, B. H. 1985. Pooling cross-sections with unequal time-series lengths. *Economics Letters* 18: 133–136.

——. 2001. *Econometric Analysis of Panel Data*. 2d ed. New York: John Wiley & Sons.

Baltagi, B. H. and Y. Chang. 1994. Incomplete Panels: A comparative study of alternative estimators for the unbalanced one-way error component regression model. *Journal of Econometrics* 62: 67–89.

Baltagi, B. H. and Q. Li. 1990. A Lagrange multiplier test for the error components model with incomplete panels. *Econometric Reviews* 9(1): 103–107.

Baum, C. F. 2001. Residual diagnostics. *The Stata Journal* 1: 101–104.

Breusch, T. and A. Pagan. 1980. The Lagrange multiplier test and its applications to model specification in econometrics. *Review of Economic Studies* 47: 239–253.

Dwyer, J. and M. Feinleib. 1992. Introduction to statistical models for longitudinal observation. In *Statistical Models for Longitudinal Studies of Health*, ed. J. Dwyer, M. Feinleib, P. Lippert, and H. Hoffmeister, 3–48. New York: Oxford University Press.

Greene, W. H. 1983. Simultaneous estimation of factor substitution, economies of scale, and non-neutral technical change. In *Econometric Analyses of Productivity*, ed. A. Dogramaci. Boston: Kluwer-Nijhoff.

——. 2003. *Econometric Analysis*. 5th ed. Upper Saddle River, NJ: Prentice–Hall.

Hausman, J. A. 1978. Specification tests in econometrics. *Econometrica* 46: 1251–1271.

Judge, G. G., W. E. Griffiths, R. C. Hill, H. Lütkepohl, and T.-C. Lee. 1985. *The Theory and Practice of Econometrics*. 2d ed. New York: John Wiley & Sons.

Lee, L. and W. Griffiths. 1979. The prior likelihood and best linear unbiased prediction in stochastic coefficient linear models. University of New England Working Papers in Econometrics and Applied Statistics No. 1, Armidale, Australia.

Rabe-Hesketh, S., A. Pickles, and C. Taylor. 2000. sg129: Generalized linear latent and mixed models. *Stata Technical Bulletin* 53: 47–57. Reprinted in *Stata Technical Bulletin Reprints*, vol. 9, pp. 293–307.

Sosa-Escudero, W. and A. K. Bera. 2001. sg164: Specification tests for linear panel data models. *Stata Technical Bulletin* 61: 18–21. Reprinted in *Stata Technical Bulletin Reprints*, vol. 10, pp. 307–311.

Swamy, P. A. V. B. and S. S. Arora. 1972. The exact finite sample properties of the estimators of coefficients in the error components regression models. *Econometrica* 40: 643–657.

Taub, A. J. 1979. Prediction in the context of the variance-components model. *Journal of Econometrics* 10: 103–108.

Wooldridge, J. M. 2002. *Econometric Analysis of Cross Section and Panel Data*. Cambridge, MA: The MIT Press.

Also See

Complementary:	[XT] **xtdata**, [XT] **xtdes**, [XT] **xtsum**, [XT] **xttab**, [R] **adjust**, [R] **lincom**, [R] **mfx**, [R] **nlcom**, [R] **predict**, [R] **predictnl**, [R] **test**, [R] **testnl**, [R] **vce**
Related:	[XT] **xtgee**, [XT] **xtintreg**, [XT] **xtivreg**, [XT] **xtregar**, [XT] **xttobit**
Background:	[U] **16.5 Accessing coefficients and standard errors**, [U] **23 Estimation and post-estimation commands**, [U] **23.14 Obtaining robust variance estimates**, [XT] **xt**

Title

> **xtregar** — Fixed- and random-effects linear models with an AR(1) disturbance

Syntax

Random-effects model

> xtregar *depvar* [*varlist*] [if *exp*] [in *range*] [, re <u>rho</u>type(*rhomethod*) <u>two</u>step
>
> rhof(*#*) lbi <u>level</u>(*#*)]

Fixed-effects model

> xtregar *depvar* [*varlist*] [*weight*] [if *exp*] [in *range*] , fe [<u>rho</u>type(*rhomethod*)
>
> <u>two</u>step rhof(*#*) lbi <u>level</u>(*#*)]

by ... : may be used with xtregar; see [R] **by**.

You must tsset your data before using xtregar; see [TS] **tsset**.

varlist may contain time-series operators; see [U] **14.4.3 Time-series varlists**.

fweights and aweights are allowed for the fixed-effects model with rhotype(regress) or rhotype(freg), or with a fixed rho. see [U] **14.1.6 weight**.

xtregar shares the features of all estimation commands; see [U] **23 Estimation and post-estimation commands**.

Syntax for predict

> predict [*type*] *newvarname* [if *exp*] [in *range*] [, *statistic*]

where *statistic* is

xb	$\mathbf{x}_{it}\mathbf{b}$,	fitted values (the default)
ue	$\nu_i + e_{it}$,	the combined residual
*u	ν_i,	the fixed component
*e	e_{it},	the overall error component

u and e are available only for the fixed-effects estimator. Unstarred statistics are available both in and out of sample; type predict ... if e(sample) ... if wanted only for the estimation sample. Starred statistics are calculated only for the estimation sample even when if e(sample) is not specified.

Description

xtregar fits cross-sectional time-series regression models when the disturbance term is first-order autoregressive. xtregar offers a within estimator for fixed-effects models and a GLS estimator for random-effects models. Consider the model

$$y_{it} = \alpha + \mathbf{x}_{it}\boldsymbol{\beta} + \nu_i + \epsilon_{it} \qquad i = 1,\ldots,N; \quad t = 1,\ldots,T_i, \tag{1}$$

where

$$\epsilon_{it} = \rho\epsilon_{i,t-1} + \eta_{it} \tag{2}$$

and where $|\rho| < 1$ and η_{it} is independent and identically distributed (i.i.d.) with mean 0 and variance σ_η^2. If ν_i are assumed to be fixed parameters, then the model is a fixed-effects model. If ν_i are assumed to be realizations of an i.i.d. process with mean 0 and variance σ_ν^2, then it is a random-effects model. Whereas in the fixed-effects model, the ν_i may be correlated with the covariates $\mathbf{x}_{it}$, in the random-effects model it is assumed that the ν_i are independent of the $\mathbf{x}_{it}$. On the other hand, any $\mathbf{x}_{it}$ that do not vary over t are collinear with the ν_i and will be dropped from the fixed-effects model. In contrast, the random-effects model can accommodate covariates that are constant over time.

xtregar can accommodate unbalanced panels whose observations are unequally spaced over time. xtregar implements the methods derived in Baltagi and Wu (1999).

Since xtregar uses time-series methods, you must tsset your data before using xtregar. See [TS] **tsset** for details.

Options

re requests the GLS estimator of the random-effects model. This is the default if neither re or fe are specified.

fe requests the within estimator of the fixed-effects model.

rhotype(*rhomethod*) allows the user to specify any of the following estimators of ρ:

dw	$\rho_{\mathrm{dw}} = 1 - d/2$, where d is the Durbin–Watson d statistic
regress	$\rho_{\mathrm{reg}} = \beta$ from the residual regression $\epsilon_t = \beta\epsilon_{t-1}$
freg	$\rho_{\mathrm{freg}} = \beta$ from the residual regression $\epsilon_t = \beta\epsilon_{t+1}$
tscorr	$\rho_{\mathrm{tscorr}} = \epsilon'\epsilon_{t-1}/\epsilon'\epsilon$, where ϵ is the vector of residuals and ϵ_{t-1} is the vector of lagged residuals
theil	$\rho_{\mathrm{theil}} = \rho_{\mathrm{tscorr}}(N - k)/N$
nagar	$\rho_{\mathrm{nagar}} = (\rho_{\mathrm{dw}}N^2 + k^2)/(N^2 - k^2)$
onestep	$\rho_{\mathrm{onestep}} = (n/m_c)(\epsilon'\epsilon_{t-1}/\epsilon'\epsilon)$, where ϵ is the vector of residuals, n is the number of observations, and m_c is the number of consecutive pairs of residuals

dw is the default method. Except for onestep, the details of these methods are given in [TS] **prais**. prais handles unequally spaced data. onestep is the one-step method proposed by Baltagi and Wu (1999). Further details on this method are available below in *Methods and Formulas*.

twostep requests that a two-step implementation of the *rhomethod* estimator of ρ be used. Unless a fixed value of ρ is specified, ρ is estimated by running prais (*sic*) on the de-meaned data. When twostep is specified, prais will stop on the first iteration after the equation is transformed by ρ—the two-step efficient estimator. Although it is customary to iterate these estimators to convergence, they are efficient at each step. When twostep is not specified, the FGLS process iterates to convergence as described in [TS] **prais**.

rhof(#) specifies that the given number is to be used for ρ and that ρ is not to be estimated.

lbi requests that the Baltagi–Wu (1999) locally best invariant (LBI) test statistic that $\rho = 0$ and a modified version of the Bhargava et al. (1982) Durbin–Watson statistic be calculated and reported. The default is not to report them. p-values are not reported for either statistic. While Bhargava et al. (1982) published critical values for their statistic, no tables are currently available for the Baltagi–Wu (LBI). Baltagi and Wu (1999) did derive a normalized version of their statistic, but this statistic cannot be computed for datasets of moderate size. One can also specify these options upon replay.

level(#) specifies the confidence level, in percent, for confidence intervals. The default is level(95) or as set by set level; see [U] **23.6 Specifying the width of confidence intervals**.

Options for predict

xb, the default, calculates the linear prediction, $\mathbf{x}_{it}\beta$.

ue calculates the prediction of $\nu_i + e_{it}$.

u calculates the prediction of ν_i, the estimated fixed effect.

e calculates the prediction of e_{it}.

Remarks

Remarks are presented under the headings

Introduction
The fixed-effects model
The random-effects model

Introduction

If you have not read [XT] **xt**, please do so.

Consider a linear panel-data model described by (1) and (2). In the fixed-effects model, the ν_i are a set of fixed parameters to be estimated. Alternatively, the ν_i may be random and correlated with the other covariates, with inference conditional on the ν_i in the sample. See Mundlak (1978) and Hsiao (1986) for a discussion of this interpretation. In the random-effects model, also known as the variance-components model, the ν_i are assumed to be realizations of an i.i.d. process with mean 0 and variance σ_ν^2. xtregar offers a within estimator for the fixed-effect model and the Baltagi–Wu (1999) GLS estimator of the random-effects model. The Baltagi–Wu (1999) GLS estimator extends the balanced panel estimator in Baltagi and Li (1991) to a case of exogenously unbalanced panels with unequally spaced observations. Both of these estimators offer several estimators of ρ.

The data can be unbalanced and unequally spaced. Specifically, the dataset contains observations on individual i at times t_{ij} for $j = 1, \ldots, n_i$. The difference $t_{ij} - t_{i,j-1}$ plays an integral role in the estimation techniques employed by xtregar. For this reason, you must tsset your data before using xtregar. For instance, if you have quarterly data, the "time" difference between the third and fourth quarter must be one month, not three.

The fixed-effects model

Let's examine the fixed-effect model first. The basic approach is common to all fixed-effects models. The ν_i are treated as nuisance parameters. We use a transformation of the model that removes the nuisance parameters and leaves behind the parameters of interest in an estimable form. Note that subtracting the group means from (1) removes the ν_i from the model

$$y_{it_{ij}} - \overline{y}_i = \left(\overline{\mathbf{x}}_{it_{ij}} - \overline{\mathbf{x}}_i\right)\beta + \epsilon_{it_{ij}} - \overline{\epsilon}_i \tag{3}$$

where

$$\overline{y}_i = \left(\frac{1}{n_i}\right)\sum_{j=1}^{n_i} y_{it_{ij}} \qquad \overline{\mathbf{x}}_i = \left(\frac{1}{n_i}\right)\sum_{j=1}^{n_i} \mathbf{x}_{it_{ij}} \qquad \overline{\epsilon}_i = \left(\frac{1}{n_i}\right)\sum_{j=1}^{n_i} \epsilon_{it_{ij}}$$

After the transformation, (3) is a linear AR(1) model, potentially with unequally spaced observations. (3) can be used to estimate ρ. Given an estimate of ρ, one must do a Cochrane–Orcutt transformation on each panel and then remove the within-panel means and add back the overall mean for each variable. OLS on the transformed data will produce the within estimates of α and β.

▷ Example

Let's use the Grunfeld investment dataset to illustrate some aspects of how xtregar can be used to fit the fixed-effects model. This dataset contains information on 10 firms' investment, market value, and the value of their capital stocks. The data were collected annually between 1935 and 1954. The following output shows that we have tsset our data, and gives the results of running a fixed-effects model with investment as a function of market value and the capital stock.

```
. use http://www.stata-press.com/data/r8/grunfeld, clear
. tsset
       panel variable:  company, 1 to 10
        time variable:  year, 1935 to 1954
. xtregar invest mvalue kstock, fe
```

FE (within) regression with AR(1) disturbances				Number of obs	=	190
Group variable (i): company				Number of groups	=	10

R-sq: within = 0.5927				Obs per group: min =	19
between = 0.7989				avg =	19.0
overall = 0.7904				max =	19

				F(2,178)	=	129.49
corr(u_i, Xb) = -0.0454				Prob > F	=	0.0000

invest	Coef.	Std. Err.	t	P>|t|	[95% Conf. Interval]	
mvalue	.0949999	.0091377	10.40	0.000	.0769677	.113032
kstock	.350161	.0293747	11.92	0.000	.2921935	.4081286
_cons	-63.22022	5.648271	-11.19	0.000	-74.36641	-52.07402

rho_ar	.67210608	
sigma_u	91.507609	
sigma_e	40.992469	
rho_fov	.8328647	(fraction of variance due to u_i)

F test that all u_i=0:   F(9,178) =   11.53   Prob > F = 0.0000

Note that since there are 10 groups, the panel-by-panel Cochrane–Orcutt method decreases the number of available observations from 200 to 190. The above example used the default dw estimator of ρ. Using the tscorr estimator of ρ yields

(Continued on next page)

```
. xtregar invest mvalue kstock, fe rhotype(tscorr)
```

```
FE (within) regression with AR(1) disturbances    Number of obs     =      190
Group variable (i): company                       Number of groups  =       10

R-sq:  within  = 0.6583                            Obs per group: min =       19
       between = 0.8024                                           avg =     19.0
       overall = 0.7933                                           max =       19

                                                  F(2,178)          =   171.47
corr(u_i, Xb)  = -0.0709                           Prob > F          =   0.0000
```

invest	Coef.	Std. Err.	t	P>|t|	[95% Conf. Interval]	
mvalue	.0978364	.0096786	10.11	0.000	.0787369	.1169359
kstock	.346097	.0242248	14.29	0.000	.2982922	.3939018
_cons	-61.84403	6.621354	-9.34	0.000	-74.91049	-48.77758
rho_ar	.54131231					
sigma_u	90.893572					
sigma_e	41.592151					
rho_fov	.82686297	(fraction of variance due to u_i)				

```
F test that all u_i=0:      F(9,178) =    19.73           Prob > F = 0.0000
```

◁

❑ Technical Note

The tscorr estimator of ρ is bounded in $[-1, 1]$. The other estimators of ρ are not. In samples with very short panels, the estimates of ρ produced by the other estimators of ρ may be outside of $[-1, 1]$. If this happens, use the tscorr estimator. However, simulations have shown that the tscorr estimator is biased toward zero. dw is the default because it performed well in Monte Carlo simulations. Note that in the example above, the estimate of ρ produced by tscorr is much smaller than the one produced by dw.

❑

▷ Example

xtregar will complain if you try to run xtregar on a dataset that has not been tsset:

```
. tsset, clear
. xtregar invest mvalue kstock, fe
must tsset data and specify panelvar
r(459);
```

xtregar requires you to tsset your data to ensure that xtregar understands the nature of your time variable. For instance, suppose that our observations were taken quarterly instead of annually. We will get exactly the same results with the quarterly variable t2 that we did with the annual variable year.

```
. generate t = year - 1934
. generate t2 = q(1934q4) + t
. format t2 %tq
```

(Continued on next page)

```
. list year t2 in 1/5
```

	year	t2
1.	1935	1935q1
2.	1936	1935q2
3.	1937	1935q3
4.	1938	1935q4
5.	1939	1936q1

```
. tsset company t2
       panel variable:  company, 1 to 10
        time variable:  t2, 1935q1 to 1939q4
. xtregar invest mvalue kstock, fe
```

```
FE (within) regression with AR(1) disturbances    Number of obs     =        190
Group variable (i): company                       Number of groups  =         10

R-sq:  within  = 0.5927                            Obs per group: min =         19
       between = 0.7989                                           avg =       19.0
       overall = 0.7904                                           max =         19

                                                   F(2,178)          =     129.49
corr(u_i, Xb)   = -0.0454                           Prob > F          =     0.0000
```

invest	Coef.	Std. Err.	t	P>|t|	[95% Conf. Interval]	
mvalue	.0949999	.0091377	10.40	0.000	.0769677	.113032
kstock	.350161	.0293747	11.92	0.000	.2921935	.4081286
_cons	-63.22022	5.648271	-11.19	0.000	-74.36641	-52.07402

rho_ar	.67210608	
sigma_u	91.507609	
sigma_e	40.992469	
rho_fov	.8328647	(fraction of variance due to u_i)

```
F test that all u_i=0:      F(9,178) =     11.53                 Prob > F = 0.0000
```

◁

In all the examples thus far, we have assumed that ϵ_{it} is first-order autoregressive. Testing the hypothesis of $\rho = 0$ in a first-order autoregressive process has a long history of producing test statistics with extremely complicated distributions. The extensions of these tests to panel data have continued this tradition. Bhargava et al. (1982) extended the Durbin–Watson statistic to the case of balanced, equally spaced panel datasets. Baltagi and Wu (1999) modified their statistic to account for unbalanced panels with unequally spaced data. In the same article, Baltagi and Wu (1999) derived the locally best invariant test statistic of $\rho = 0$. Both these test statistics have extremely complicated distributions, although Bhargava et al. (1982) did publish some cutoffs in their article. Specifying the lbi option to xtregar will cause Stata to calculate and report the modified Bhargava et al. Durbin–Watson and the Baltagi–Wu LBI.

▷ Example

In this example, we calculate the modified Bhargava et al. Durbin–Watson statistic and the Baltagi–Wu LBI. We exclude time periods 9 and 10 from the sample, thereby reproducing the results of Baltagi and Wu (1999, 822). Note that p-values are not reported for either statistic. While Bhargava et al. (1982) published critical values for their statistic, no tables are currently available for the Baltagi–Wu (LBI). Baltagi and Wu (1999) did derive a normalized version of their statistic, but this statistic cannot be computed for datasets of moderate size.

```
. xtregar invest mvalue kstock if year !=1934 & year !=1944, fe lbi

FE (within) regression with AR(1) disturbances    Number of obs      =      180
Group variable (i): company                       Number of groups   =       10

R-sq:  within  = 0.5954                            Obs per group: min =       18
       between = 0.7952                                           avg =     18.0
       overall = 0.7889                                           max =       18

                                                   F(2,168)           =   123.63
corr(u_i, Xb)  = -0.0516                            Prob > F           =   0.0000
```

invest	Coef.	Std. Err.	t	P>|t|	[95% Conf. Interval]	
mvalue	.0941122	.0090926	10.35	0.000	.0761617	.1120627
kstock	.3535872	.0303562	11.65	0.000	.2936584	.4135161
_cons	-64.82534	5.946885	-10.90	0.000	-76.56559	-53.08509

rho_ar	.6697198	
sigma_u	93.320452	
sigma_e	41.580712	
rho_fov	.83435413	(fraction of variance due to u_i)

```
F test that all u_i=0:     F(9,168) =     11.55          Prob > F = 0.0000
modified Bhargava et al. Durbin-Watson = .71380994
Baltagi-Wu LBI = 1.0134522
```

◁

The random-effects model

In the random-effects model, the ν_i are assumed to be realizations of an i.i.d. process with mean 0 and variance σ_ν^2. Furthermore, the ν_i are assumed to be independent of both the ϵ_{it} and the covariates x_{it}. The latter of these assumptions can be very strong. However, inference is not conditional on the particular realizations of the ν_i in the sample. See Mundlak (1978) for a discussion of this point.

▷ Example

By specifying the re option, one obtains the Baltagi–Wu GLS estimator of the random-effects model. This estimator can accommodate unbalanced panels and unequally spaced data. We run this model on the Grunfeld dataset:

(Continued on next page)

```
. xtregar invest mvalue kstock if year!=1934 & year !=1944, re lbi
RE GLS regression with AR(1) disturbances       Number of obs      =        190
Group variable (i): company                     Number of groups   =         10
R-sq:  within  = 0.7707                          Obs per group: min =         19
       between = 0.8039                                         avg =       19.0
       overall = 0.7958                                         max =         19

                                                 Wald chi2(3)       =     351.37
corr(u_i, Xb)       = 0 (assumed)                Prob > chi2        =     0.0000
```

invest	Coef.	Std. Err.	z	P>|z|	[95% Conf. Interval]
mvalue	.0947714	.0083691	11.32	0.000	.0783683 .1111746
kstock	.3223932	.0263226	12.25	0.000	.2708019 .3739845
_cons	-45.21427	27.12492	-1.67	0.096	-98.37814 7.949603

rho_ar	.6697198	(estimated autocorrelation coefficient)
sigma_u	74.662876	
sigma_e	42.253042	
rho_fov	.75742494	(fraction of variance due to u_i)
theta	.66973313	

```
modified Bhargava et al. Durbin-Watson = .71380994
Baltagi-Wu LBI = 1.0134522
```

Note that the modified Bhargava et al. Durbin–Watson and the Baltagi–Wu LBI are exactly the same as those reported for the fixed-effects model because the formulas for these statistics do not depend on whether we are fitting the fixed-effects model or the random-effects model.

◁

(Continued on next page)

Saved Results

xtregar, re saves in e():

Scalars

e(d1)	Bhargava et al. Durbin–Watson	e(LBI)	Baltagi–Wu LBI statistic
e(ds)	centered Baltagi–Wu LBI	e(N_LBI)	number of obs used in e(LBI)
e(N)	number of observations	e(r2_o)	R-squared for overall model
e(N_g)	number of groups	e(r2_b)	R-squared for between model
e(df_m)	model degrees of freedom	e(rho_ar)	autocorrelation coefficient
e(g_max)	largest group size	e(sigma_u)	panel-level standard deviation
e(g_min)	smallest group size	e(sigma_e)	standard deviation of ϵ_{it}
e(g_avg)	average group size	e(thta_min)	minimum θ
e(chi2)	χ^2	e(thta_5)	θ, 5th percentile
e(rho_fov)	u_i fraction of variance	e(thta_50)	θ, 50th percentile
e(Tbar)	harmonic mean of group sizes	e(thta_95)	θ, 95th percentile
e(Tcon)	1 if T is constant	e(thta_max)	maximum θ
e(r2_w)	R-squared for within model		

Macros

e(cmd)	xtregar	e(ivar)	variable denoting groups
e(depvar)	name of dependent variable	e(tvar)	time variable
e(model)	re	e(chi2type)	Wald; type of model χ^2 test
e(rhotype)	method of estimating ρ_{ar}	e(predict)	program used to implement predict
e(dw)	LBI, if requested		

Matrices

e(b)	coefficient vector	e(V)	VCE for random-effects model

Functions

e(sample)	marks estimation sample

(Continued on next page)

`xtregar, fe` saves in `e()`:

Scalars

e(d1)	Bhargava et al. Durbin–Watson	e(LBI)	Baltagi–Wu LBI statistic
e(ds)	centered Baltagi–Wu LBI	e(N_LBI)	number of obs used in e(LBI)
e(N)	number of observations	e(g_max)	largest group size
e(N_g)	number of groups	e(g_min)	smallest group size
e(mss)	model sum of squares	e(g_avg)	average group size
e(tss)	total sum of squares	e(rho_fov)	u_i fraction of variance
e(df_m)	model degrees of freedom	e(Tbar)	harmonic mean of group sizes
e(rss)	residual sum of squares	e(Tcon)	1 if T is constant
e(df_r)	residual degrees of freedom	e(r2_w)	R-squared for within model
e(r2)	R-squared	e(r2_o)	R-squared for overall model
e(r2_a)	adjusted R-squared	e(r2_b)	R-squared for between model
e(F)	F statistic	e(rho_ar)	autocorrelation coefficient
e(rmse)	root mean square error	e(corr)	corr(u_i, Xb)
e(ll)	log likelihood	e(sigma_u)	panel-level standard deviation
e(ll_0)	log likelihood, constant-only model	e(sigma_e)	standard deviation of ϵ_{it}
e(df_a)	degrees of freedom for absorbed effect	e(F_f)	F for $u_i=0$

Macros

e(cmd)	xtregar	e(ivar)	variable denoting groups
e(depvar)	name of dependent variable	e(tvar)	time variable
e(model)	fe	e(wtype)	weight type
e(rhotype)	method of estimating ρ_{ar}	e(wexp)	weight expression
e(dw)	LBI, if requested	e(predict)	program used to implement predict

Matrices

e(b)	coefficient vector	e(V)	variance–covariance matrix of the estimators

Functions

e(sample)	marks estimation sample

Methods and Formulas

Consider a linear panel-data model described by (1) and (2). The data can be unbalanced and unequally spaced. Specifically, the dataset contains observations on individual i at times t_{ij} for $j = 1, \ldots, n_i$.

Estimating ρ

The estimate of ρ is always obtained after removing the group means. Let $\widetilde{y}_{it} = y_{it} - \overline{y}_i$, let $\widetilde{\mathbf{x}}_{it} = \mathbf{x}_{it} - \overline{\mathbf{x}}_i$, and let $\widetilde{\epsilon}_{it} = \epsilon_{it} - \overline{\epsilon}_i$.

Then, except for the `onestep` method, all the estimates of ρ are obtained by running Stata's `prais` on

$$\widetilde{y}_{it} = \widetilde{x}_{it}\boldsymbol{\beta} + \widetilde{\epsilon}_{it}$$

See [TS] **prais** for the formulas for each of the methods.

When `onestep` is specified, a regression is run on the above equation and the residuals are obtained. Let $e_{it_{ij}}$ be the residual used to estimate the error $\widetilde{\epsilon}_{it_{ij}}$. If $t_{ij} - t_{i,j-1} > 1$, then $e_{it_{ij}}$ is set to zero. Given this series of residuals,

$$\widehat{\rho}_{\text{onestep}} = \frac{n}{m_c} \frac{\sum_{i=1}^{N} \sum_{t=2}^{T} e_{it} e_{i,t-1}}{\sum_{i=1}^{N} \sum_{t=1}^{T} e_{it}^2}$$

where n is the number of nonzero elements in e and m_c is the number of consecutive pairs of nonzero e_{it}s.

Transforming the data to remove the AR(1) component

After estimating ρ, Baltagi and Wu (1999) derive a transformation of the data that removes the AR(1) component. Their $C_i(\rho)$ can be written as

$$y_{it_{ij}}^* = \begin{cases} (1-\rho^2)^{\frac{1}{2}} y_{it_{ij}} & \text{if } t_{ij} = 1 \\ (1-\rho^2)^{\frac{1}{2}} \left[y_{i,t_{ij}} \left\{ \frac{1}{1-\rho^{2(t_{ij}-t_{i,j-1})}} \right\}^{\frac{1}{2}} - y_{i,t_{i,j-1}} \left\{ \frac{\rho^{2(t_{ij}-t_{i,j-1})}}{1-\rho^{2(t_{i,j}-t_{i,j-1})}} \right\}^{\frac{1}{2}} \right] & \text{if } t_{ij} > 1 \end{cases}$$

Using the analogous transform on the independent variables generates transformed data that are cleansed of the AR(1) process. Performing simple OLS on the transformed data leaves behind the residuals μ^*.

The within estimator of the fixed-effects model

To obtain the within estimator, we need to transform the data that come out of the AR(1) transform. In order for the within transform to remove the fixed-effects, the first observation of each panel must be dropped. Specifically, let

$$\breve{y}_{it_{ij}} = y_{it_{ij}}^* - \overline{y}_i^* + \overline{\overline{y}}^* \qquad \forall j > 1$$
$$\breve{\mathbf{x}}_{it_{ij}} = \mathbf{x}_{it_{ij}}^* - \overline{\mathbf{x}}_i^* + \overline{\overline{\mathbf{x}}}^* \qquad \forall j > 1$$
$$\breve{\epsilon}_{it_{ij}} = \epsilon_{it_{ij}}^* - \overline{\epsilon}_i^* + \overline{\overline{\epsilon}}^* \qquad \forall j > 1$$

where

$$\overline{y}_i^* = \frac{\sum_{j=2}^{n_i-1} y_{it_{ij}}^*}{n_i - 1}$$

$$\overline{\overline{y}}^* = \frac{\sum_{i=1}^{N} \sum_{j=2}^{n_i-1} y_{it_{ij}}^*}{\sum_{i=1}^{N} n_i - 1}$$

$$\overline{\mathbf{x}}_i^* = \frac{\sum_{j=2}^{n_i-1} \mathbf{x}_{it_{ij}}^*}{n_i - 1}$$

$$\overline{\overline{\mathbf{x}}}^* = \frac{\sum_{i=1}^{N} \sum_{j=2}^{n_i-1} \mathbf{x}_{it_{ij}}^*}{\sum_{i=1}^{N} n_i - 1}$$

$$\overline{\epsilon}_i^* = \frac{\sum_{j=2}^{n_i-1} \epsilon_{it_{ij}}^*}{n_i - 1}$$

$$\overline{\overline{\epsilon}}^* = \frac{\sum_{i=1}^{N} \sum_{j=2}^{n_i-1} \epsilon_{it_{ij}}^*}{\sum_{i=1}^{N} n_i - 1}$$

The within estimator of the fixed-effects model is then obtained by running OLS on

$$\breve{y}_{it_{ij}} = \alpha + \breve{\mathbf{x}}_{it_{ij}} \boldsymbol{\beta} + \breve{\epsilon}_{it_{ij}}$$

Reported as R^2 within is the R^2 from the above regression.

Reported as R^2 between is $\left\{ \mathrm{corr}(\overline{\mathbf{x}}_i \widehat{\boldsymbol{\beta}}, \overline{y}_i) \right\}^2$.

Reported as R^2 overall is $\left\{ \mathrm{corr}(\mathbf{x}_{it} \widehat{\boldsymbol{\beta}}, y_{it}) \right\}^2$.

The Baltagi–Wu GLS estimator

The residuals μ^* can be used to estimate the variance components. Translating the matrix formulas given in Baltagi and Wu (1999) into summations yields the following variance-components estimators:

$$\widehat{\sigma}_\omega^2 = \sum_{i=1}^{N} \frac{(\mu_i^{*\prime} g_i)^2}{(g_i' g_i)}$$

$$\widehat{\sigma}_\epsilon^2 = \frac{\left[\sum_{i=1}^{N} (\mu_i^{*\prime} \mu_i^*) - \sum_{i=1}^{N} \left\{ \frac{(\mu_i^{*\prime} g_i)^2}{(g_i' g_i)} \right\} \right]}{\sum_{i=1}^{N} (n_i - 1)}$$

$$\widehat{\sigma}_\mu^2 = \frac{\left[\sum_{i=1}^{N} \left\{ \frac{(\mu_i^{*\prime} g_i)^2}{(g_i' g_i)} \right\} - N \widehat{\sigma}_\epsilon^2 \right]}{\sum_{i=1}^{N} (g_i' g_i)}$$

where

$$g_i = \left[1, \frac{\left\{ 1 - \rho^{(t_{i,2} - t_{i,1})} \right\}}{\left\{ 1 - \rho^{2(t_{i,2} - t_{i,1})} \right\}^{\frac{1}{2}}}, \dots, \frac{\left\{ 1 - \rho^{(t_{i,n_i} - t_{i,n_i-1})} \right\}}{\left\{ 1 - \rho^{2(t_{i,n_i} - t_{i,n_i-1})} \right\}^{\frac{1}{2}}} \right]'$$

and μ_i^* is the $n_i \times 1$ vector of residuals from μ^* that correspond to person i.

Then,

$$\widehat{\theta}_i = 1 - \left(\frac{\widehat{\sigma}_\mu}{\widehat{\omega}_i} \right)$$

where

$$\widehat{\omega}_i^2 = g_i' g_i \widehat{\sigma}_\mu^2 + \widehat{\sigma}_\epsilon^2$$

With these estimates in hand, one can transform the data via

$$z_{it_{ij}}^{**} = z_{it_{ij}}^* - \widehat{\theta}_i g_{ij} \frac{\sum_{s=1}^{n_i} g_{is} z_{it_{is}}^*}{\sum_{s=1}^{n_i} g_{is}^2}$$

for $z \in \{y, \mathbf{x}\}$.

Running OLS on the transformed data $y^{**}, \mathbf{x}^{**}$ yields the feasible GLS estimator of α and $\boldsymbol{\beta}$.

Reported as R^2 between is $\left\{ \mathrm{corr}(\overline{\mathbf{x}}_i \widehat{\boldsymbol{\beta}}, \overline{y}_i) \right\}^2$.

Reported as R^2 within is $\left\{ \mathrm{corr}\{ (\mathbf{x}_{it} - \overline{\mathbf{x}}_i) \widehat{\boldsymbol{\beta}}, y_{it} - \overline{y}_i \} \right\}^2$.

Reported as R^2 overall is $\left\{ \mathrm{corr}(\mathbf{x}_{it} \widehat{\boldsymbol{\beta}}, y_{it}) \right\}^2$.

The test statistics

The Baltagi–Wu LBI is the sum of terms

$$d_* = d_1 + d_2 + d_3 + d_4$$

where

$$d_1 = \frac{\sum_{i=1}^{N} \sum_{j=1}^{n_i} \{ \widetilde{z}_{it_{i,j-1}} - \widetilde{z}_{it_{ij}} I(t_{ij} - t_{i,j-1} = 1) \}^2}{\sum_{i=1}^{N} \sum_{j=1}^{n_i} \widetilde{z}_{it_{ij}}^2}$$

$$d_2 = \frac{\sum_{i=1}^{N} \sum_{j=1}^{n_i - 1} \widetilde{z}_{it_{i,j-1}}^2 \{ 1 - I(t_{ij} - t_{i,j-1} = 1) \}^2}{\sum_{i=1}^{N} \sum_{j=1}^{n_i} \widetilde{z}_{it_{ij}}^2}$$

$$d_3 = \frac{\sum_{i=1}^{N} \widetilde{z}_{it_{i1}}^2}{\sum_{i=1}^{N} \sum_{j=1}^{n_i} \widetilde{z}_{it_{ij}}^2}$$

$$d_4 = \frac{\sum_{i=1}^{N} \widetilde{z}_{it_{in_i}}^2}{\sum_{i=1}^{N} \sum_{j=1}^{n_i} \widetilde{z}_{it_{ij}}^2}$$

$I()$ is the indicator function that takes the value of 1 if the condition is true and 0 otherwise. The $\widetilde{z}_{it_{i,j-1}}$ are residuals from the within estimator.

Baltagi and Wu (1999) also show that d_1 is the Bhargava et al. Durbin–Watson statistic modified to handle cases of unbalanced panels and unequally spaced data.

Acknowledgment

We would like to thank Badi Baltagi, Department of Economics, Texas A&M University for his helpful comments.

References

Baltagi, B. H. 2001. *Econometric Analysis of Panel Data.* 2d ed. New York: John Wiley & Sons.

Baltagi, B. H. and Q. Li. 1991. A transformation that will circumvent the problem of autocorrelation in an error component model. *Journal of Econometrics* 48: 385–393.

Baltagi, B. H. and P. X. Wu. 1999. Unequally spaced panel data regressions with AR(1) disturbances. *Econometric Theory* 15: 814–823.

Bhargava, A., L. Franzini, and W. Narendranathan. 1982. Serial correlation and the fixed effects model. *The Review of Economic Studies* 49: 533–549.

Hsiao, C. 1986. *Analysis of Panel Data.* New York: Cambridge University Press.

Mundlak, Y. 1978. On the pooling of time series and cross section data. *Econometrica* 46: 69–85.

Also See

Complementary:	[XT] **xtdata**, [XT] **xtdes**, [XT] **xtsum**, [XT] **xttab**,
	[R] **adjust**, [R] **lincom**, [R] **mfx**, [R] **nlcom**, [R] **predict**,
	[R] **predictnl**, [R] **test**, [R] **testnl**, [R] **vce**,
	[TS] **tsset**
Related:	[XT] **xtgee**, [XT] **xtintreg**, [XT] **xtivreg**, [XT] **xtreg**, [XT] **xttobit**
Background:	[U] **16.5 Accessing coefficients and standard errors**,
	[U] **23 Estimation and post-estimation commands**,
	[XT] **xt**

Title

> **xtsum** — Summarize xt data

Syntax

xtsum [*varlist*] [if *exp*] [, i(*varname*)]

by ... : may be used with xtsum; see [R] **by**.

Description

xtsum, a generalization of summarize, reports means and standard deviations for cross-sectional time-series (xt) data; it differs from summarize in that it decomposes the standard deviation into between and within components.

Options

i(*varname*) specifies the variable name that contains the unit to which the observation belongs. You can specify the i() option the first time you estimate, or you can use the iis command to set i() beforehand. Note that it is not necessary to specify i() if the data have been previously tsset, or if iis has been previously specified—in these cases, the group variable is taken from the previous setting. See [XT] **xt**.

Remarks

If you have not read [XT] **xt**, please do so.

xtsum provides an alternative to summarize. For instance, in the nlswork dataset described in [XT] **xt**, hours contains the number of hours worked last week:

```
. use http://www.stata-press.com/data/r8/nlswork
(National Longitudinal Survey.  Young Women 14-26 years of age in 1968)

. summarize hours
```

Variable	Obs	Mean	Std. Dev.	Min	Max
hours	28467	36.55956	9.869623	1	168

```
. xtsum hours, i(idcode)
```

Variable		Mean	Std. Dev.	Min	Max	Observations
hours	overall	36.55956	9.869623	1	168	N = 28467
	between		7.846585	1	83.5	n = 4710
	within		7.520712	-2.154726	130.0596	T-bar = 6.04395

xtsum provides the same information as summarize and more. It decomposes the variable x_{it} into a between ($\bar{x}_i$) and within ($x_{it} - \bar{x}_i + \bar{\bar{x}}$; the global mean $\bar{\bar{x}}$ being added back in make results comparable). The overall and within are calculated over 28,467 person-years of data. The between is calculated over 4,710 persons. And, for your information, the average number of years a person was observed in the hours data is 6.

xtsum also reports minimums and maximums: Hours worked last week varied between 1 and (unbelievably) 168. Average hours worked last week for each woman varied between 1 and 83.5. "Hours worked within" varied between -2.15 and 130.1, which is not to say any woman actually worked negative hours. The within number refers to the deviation from each individual's average, and naturally, some of those deviations must be negative. In that case, it is not the negative value that is disturbing but the positive value. Did some woman really deviate from her average by $+130.1$ hours? No; in our definition of within, we add back in the global average of 36.6 hours. Some woman did deviate from her average by $130.1 - 36.6 = 93.5$ hours, which is still quite large.

The reported standard deviations tell us something that may surprise you. They say that the variation in hours worked last week across women is very nearly equal to that observed within a woman over time. That is, if you were to draw two women randomly from our data, the difference in hours worked is expected to be nearly equal to the difference for the same woman in two randomly selected years.

If a variable does not vary over time, its within standard deviation will be zero:

```
. xtsum birth_yr
Variable          |     Mean   Std. Dev.       Min        Max |    Observations

birth_yr overall  | 48.08509   3.012837         41         54 |  N   =    28534
         between  |            3.051795         41         54 |  n   =     4711
         within   |                   0   48.08509   48.08509 |  T-bar = 6.05689
```

Also See

Related: [XT] **xtdes**, [XT] **xttab**

Background: [XT] **xt**

Title

> **xttab** — Tabulate xt data

Syntax

> xttab *varname* [if *exp*] [, i(*varname*$_i$)]
>
> xttrans *varname* [if *exp*] [, i(*varname*$_i$) t(*varname*$_t$) <u>freq</u>]

by ... : may be used with xttab and xttrans; see [R] **by**.

Description

xttab, a generalization of tabulate, performs one-way tabulations, and decomposes counts into between and within components in cross-sectional time-series (xt) data.

xttrans, another generalization of tabulate, reports transition probabilities (the change in a single categorical variable over time).

Options

i(*varname*$_i$) specifies the variable name that contains the unit to which the observation belongs. You can specify the i() option the first time you estimate, or you can use the iis command to set i() beforehand. Note that it is not necessary to specify i() if the data have been previously tsset, or if iis has been previously specified—in these cases, the group variable is taken from the previous setting. See [XT] **xt**.

t(*varname*$_t$) specifies the variable that contains the time at which the observation was made. You can specify the t() option the first time you estimate, or you can use the tis command to set t() beforehand. Note that it is not necessary to specify t() if the data have been previously tsset, or if tis has been previously specified—in these cases, the group variable is taken from the previous setting. See [XT] **xt**.

freq, allowed with xttrans only, specifies that frequencies as well as transition probabilities are to be displayed.

Remarks

If you have not read [XT] **xt**, please do so.

▷ Example

Using the nlswork dataset described in [XT] **xt**, variable msp is 1 if a woman is married and her spouse resides with her and 0 otherwise:

```
. use http://www.stata-press.com/data/r8/nlswork
(National Longitudinal Survey.  Young Women 14-26 years of age in 1968)
. xttab msp, i(idcode)
```

	Overall		Between		Within
msp	Freq.	Percent	Freq.	Percent	Percent
0	11324	39.71	3113	66.08	55.06
1	17194	60.29	3643	77.33	71.90
Total	28518	100.00	6756	143.41	64.14

(n = 4711)

The overall part of the table summarizes results in terms of person-years. We have 11,324 person-years of data in which msp is 0 and 17,194 in which it is 1—in 60.3% of our data, the woman is married with her spouse present. Between repeats the breakdown, but this time in terms of women rather than woman-years; 3,113 of our women ever had msp 0 and 3,643 ever had msp 1, for a grand total of 6,756 ever having either. We have in our data, however, only 4,711 women. This means that there are women who sometimes have msp 0 and at other times have msp 1.

The within percent tells us the fraction of the time a woman has the specified value of msp. Taking the first line, conditional on a woman ever having msp 0, 55.1% of her observations have msp 0. Similarly, conditional on a woman ever having msp 1, 71.9% of her observations have msp 1. These two numbers are a measure of the stability of the msp values, and, in fact, msp 1 is more stable among these younger women than msp 0, meaning that they tend to marry more than they divorce. The total within of 64.14 percent is the normalized between weighted average of the within percents. To wit, $(3113 \times 55.06 + 3643 \times 71.90)/6756$. It is a measure of the overall stability of the msp variable.

A time-invariant variable will have a tabulation with within percents of 100:

```
. xttab race
```

	Overall		Between		Within
race	Freq.	Percent	Freq.	Percent	Percent
1	20180	70.72	3329	70.66	100.00
2	8051	28.22	1325	28.13	100.00
3	303	1.06	57	1.21	100.00
Total	28534	100.00	4711	100.00	100.00

(n = 4711)

◁

▷ Example

xttrans shows the transition probabilities. In cross-sectional time-series data, one can estimate the probability that $x_{i,t+1} = v_2$ given that $x_{it} = v_1$ by counting transitions. For instance,

```
. xttrans msp, t(year)
```

1 if married, spouse present	1 if married, spouse present		Total
	0	1	
0	80.49	19.51	100.00
1	7.96	92.04	100.00
Total	37.11	62.89	100.00

The rows reflect the initial values and the columns reflect the final values. Each year, some 80% of the msp 0 persons in the data remained msp 0 in the next year; the remaining 20% became msp 1. While msp 0 had a 20% chance of becoming msp 1 in each year, the msp 1 had only an 8% chance of becoming (or returning to) msp 0. The freq option displays the frequencies that go into the calculation:

```
. xttrans msp, freq
      1 if |
   married, | 1 if married, spouse
    spouse  |      present
   present  |       0          1  |     Total
-----------+----------------------+-----------
         0 |    7,697      1,866  |     9,563
           |   80.49      19.51   |    100.00
-----------+----------------------+-----------
         1 |    1,133     13,100  |    14,233
           |    7.96      92.04   |    100.00
-----------+----------------------+-----------
     Total |    8,830     14,966  |    23,796
           |   37.11      62.89   |    100.00
```

◁

□ Technical Note

The transition probabilities reported by xttrans are not necessarily the transition probabilities in a Markov sense. xttrans counts transitions from each observation to the next once the observations have been put in t order within i. It does not normalize for missing time periods. xttrans does pay attention to missing values of the variable being tabulated, however, and does not count transitions from nonmissing to missing and from missing to nonmissing. Thus, if the data are fully rectangularized, xttrans does produce (inefficient) estimates of the Markov transition matrix. fillin will rectangularize datasets; see [R] **fillin**. Thus, the Markov transition matrix could be estimated by typing

```
. fillin idcode year
. xttrans msp
  (output omitted)
```

□

Also See

Related: [XT] **xtdes**, [XT] **xtsum**

Background: [XT] **xt**

Title

> **xttobit** — Random-effects tobit models

Syntax

Random-effects model

> xttobit *depvar* [*varlist*] [*weight*] [if *exp*] [in *range*] [, i(*varname*) quad(#)
>
> noconstant noskip level(#) offset(*varname*) ll(*varname* | #) ul(*varname* | #)
>
> tobit nolog *maximize_options*]

by ... : may be used with xttobit; see [R] **by**.

iweights are allowed; see [U] **14.1.6 weight**. Note that weights must be constant within panels.

xttobit shares the features of all estimation commands; see [U] **23 Estimation and post-estimation commands**.

Syntax for predict

> predict [*type*] *newvarname* [if *exp*] [in *range*] [, *statistics* nooffset]

where *statistic* is

xb	linear prediction assuming $\nu_i = 0$, the default
pr0(*a,b*)	$\Pr(a < y < b)$ assuming $\nu_i = 0$
e0(*a,b*)	$E(y \mid a < y < b)$ assuming $\nu_i = 0$
ystar0(*a,b*)	$E(y^*)$, $y^* = \max\{a, \min(y, b)\}$ assuming $\nu_i = 0$
stdp	standard error of the linear prediction
stdf	standard error of the linear forecast

where *a* and *b* may be numbers or variables; *a* missing ($a \geq .$) means $-\infty$, and *b* missing ($b \geq .$) means $+\infty$; see [U] **15.2.1 Missing values**.

These statistics are available both in and out of sample; type predict ... if e(sample) ... if wanted only for the estimation sample.

Description

xttobit fits random-effects tobit models. There is no command for a parametric conditional fixed-effects model, as there does not exist a sufficient statistic allowing the fixed effects to be conditioned out of the likelihood. It should be noted that Honoré (1992) has developed a semi-parametric estimator for fixed-effect tobit models. Unconditional fixed-effects tobit models may be fitted with the tobit command with indicator variables for the panels. The appropriate indicator variables can be generated using tabulate or xi. However, unconditional fixed-effects estimates are biased.

Note: xttobit is slow since it is calculated by quadrature; see *Methods and Formulas*. Computation time is roughly proportional to the number of points used for the quadrature. The default is quad(12). Simulations indicate that increasing it does not appreciably change the estimates for the coefficients or their standard errors. See [XT] **quadchk**.

Options

i(*varname*) specifies the variable name that contains the unit to which the observation belongs. You can specify the i() option the first time you estimate, or you can use the iis command to set i() beforehand. Note that it is not necessary to specify i() if the data have been previously tsset, or if iis has been previously specified—in these cases, the group variable is taken from the previous setting. See [XT] **xt**.

quad(#) specifies the number of points to use in the quadrature approximation of the integral.

The default is quad(12). The number specified must be an integer between 4 and 30, and must be no greater than the number of observations.

noconstant suppresses the constant term (intercept) in the model.

noskip specifies that a full maximum-likelihood model with only a constant for the regression equation be fitted. This model is not displayed, but is used as the base model to compute a likelihood-ratio test for the model test statistic displayed in the estimation header. By default, the overall model test statistic is an asymptotically equivalent Wald test of all the parameters in the regression equation being zero (except the constant). For many models, this option can substantially increase estimation time.

level(#) specifies the confidence level, in percent, for confidence intervals. The default is level(95) or as set by set level; see [U] **23.6 Specifying the width of confidence intervals**.

offset(*varname*) specifies that *varname* is to be included in the model with its coefficient constrained to be 1.

ll(*varname*|#) and ul(*varname*|#) indicate the censoring points. You may specify one or both. ll() indicates the lower limit for left censoring. Observations with $depvar \leq$ ll() are left-censored; observations with $depvar \geq$ ul() are right-censored; and remaining observations are not censored.

tobit specifies that a likelihood-ratio test comparing the random-effects model with the pooled (tobit) model should be included in the output.

nolog suppresses the iteration log.

maximize_options control the maximization process; see [R] **maximize**. Use the trace option to view parameter convergence. Use the ltol(#) option to relax the convergence criterion; the default is 1e−6 during specification searches.

(Continued on next page)

Options for predict

xb, the default, calculates the linear prediction.

pr0(a,b) calculates estimates of $\Pr(a < y < b \mid \mathbf{x} = \mathbf{x}_{it}, \nu_i = 0)$, the probability that y would be observed in the interval (a, b), given the current values of the predictors, $\mathbf{x}_{it}$, and given a zero random effect; see *Remarks*. In the discussion that follows, these two conditions are taken to be implicit.

a and b may be specified as numbers or variable names; *lb* and *ub* are variable names;
pr0(20,30) calculates $\Pr(20 < y < 30)$;
pr0(*lb,ub*) calculates $\Pr(lb < y < ub)$; and
pr0(20,*ub*) calculates $\Pr(20 < y < ub)$.

a missing ($a \geq .$) means $-\infty$; pr0(.,30) calculates $\Pr(-\infty < y < 30)$;
pr0(*lb*,30) calculates $\Pr(-\infty < y < 30)$ in observations for which $lb \geq .$
(and calculates $\Pr(lb < y < 30)$ elsewhere).

b missing ($b \geq .$) means $+\infty$; pr0(20,.) calculates $\Pr(+\infty > y > 20)$;
pr0(20,*ub*) calculates $\Pr(+\infty > y > 20)$ in observations for which $ub \geq .$
(and calculates $\Pr(20 < y < ub)$ elsewhere).

e0(a,b) calculates estimates of $E(y \mid a < y < b, \mathbf{x} = \mathbf{x}_{it}, \nu_i = 0)$, the expected value of y conditional on y being in the interval (a, b), which is to say, y is censored.
a and b are specified as they are for pr0().

ystar0(a,b) calculates estimates of $E(y^* \mid \mathbf{x} = \mathbf{x}_{it}, \nu_i = 0)$, where $y^* = a$ if $y \leq a$, $y^* = b$ if $y \geq b$, and $y^* = y$ otherwise, which is to say, y^* is the truncated version of y. a and b are specified as they are for pr0().

stdp calculates the standard error of the prediction. It can be thought of as the standard error of the predicted expected value or mean for the observation's covariate pattern. This is also referred to as the standard error of the fitted value.

stdf calculates the standard error of the forecast. This is the standard error of the point prediction for a single observation. It is commonly referred to as the standard error of the future or forecast value. By construction, the standard errors produced by stdf are always larger than those by stdp; see [R] **regress** *Methods and Formulas*.

nooffset is relevant only if you specify offset(*varname*) for xttobit. It modifies the calculations made by predict so that they ignore the offset variable; the linear prediction is treated as $\mathbf{x}_{it}\boldsymbol{\beta}$ rather than $\mathbf{x}_{it}\boldsymbol{\beta} + \text{offset}_{it}$.

Remarks

Consider the linear regression model with panel-level random effects

$$y_{it} = \mathbf{x}_{it}\boldsymbol{\beta} + \nu_i + \epsilon_{it}$$

for $i = 1, \ldots, n$, $t = 1, \ldots, n_i$. The random effects, ν_i, are iid $N(0, \sigma_\nu^2)$, and ϵ_{it} are iid $N(0, \sigma_\epsilon^2)$ independently of ν_i.

The observed data, y_{it}^o, represent possibly censored versions of y_{it}. If left-censored, then all that is known is that $y_{it} \leq y_{it}^o$. If right-censored, then all that is known is that $y_{it} \geq y_{it}^o$. If uncensored, then $y_{it} = y_{it}^o$. If left-censored, then y_{it}^o is determined by ll(). If right-censored, then y_{it}^o is determined by ul(). If uncensored, then y_{it}^o is determined by *depvar*.

▷ Example

Using the `nlswork` data described in [XT] **xt**, we fit a random-effects tobit model of adjusted (log) wages. We use the `ul()` option to impose an upper limit on the recorded log of wages.

```
. use http://www.stata-press.com/data/r8/nlswork
(National Longitudinal Survey.  Young Women 14-26 years of age)
. generate southXt = south*(year-70)
(8 missing values generated)
. xttobit ln_wage union age grade not_smsa south southXt occ_code, i(id) ul(1.9)
> tobit nolog
```

Random-effects tobit regression			Number of obs	=	19151
Group variable (i): idcode			Number of groups	=	4140

Random effects u_i ~ Gaussian	Obs per group: min =	1
	avg =	4.6
	max =	12

	Wald chi2(7)	=	3361.19
Log likelihood = -6665.7879	Prob > chi2	=	0.0000

ln_wage	Coef.	Std. Err.	z	P>|z|	[95% Conf. Interval]	
union	.1539529	.0069077	22.29	0.000	.140414	.1674918
age	.0086198	.0005423	15.89	0.000	.0075569	.0096826
grade	.0789552	.0022067	35.78	0.000	.0746303	.0832802
not_smsa	-.1269592	.00889	-14.28	0.000	-.1443834	-.109535
south	-.1182003	.0120556	-9.80	0.000	-.1418287	-.0945718
southXt	.0031335	.0008385	3.74	0.000	.0014901	.0047769
occ_code	-.018946	.0010936	-17.32	0.000	-.0210894	-.0168026
_cons	.5596519	.0332332	16.84	0.000	.494516	.6247878
/sigma_u	.2838666	.0043967	64.56	0.000	.2752492	.292484
/sigma_e	.2495757	.0018144	137.55	0.000	.2460196	.2531319
rho	.5640177	.0082238			.5478505	.5800786

```
Likelihood-ratio test of sigma_u=0: chibar2(01)= 5921.31 Prob>=chibar2 = 0.000
     Observation summary:     12288   uncensored observations
                                  0 left-censored observations
                               6863 right-censored observations
```

The output includes the overall and panel-level variance components (labeled `sigma_e` and `sigma_u`, respectively) together with ρ (labeled `rho`),

$$\rho = \frac{\sigma_\nu^2}{\sigma_\epsilon^2 + \sigma_\nu^2}$$

which is the percent contribution to the total variance of the panel-level variance component.

When `rho` is zero, the panel-level variance component is unimportant, and the panel estimator is not different from the pooled estimator. A likelihood-ratio test of this is included at the bottom of the output. This test formally compares the pooled estimator (tobit) with the panel estimator. ◁

❏ Technical Note

   The random-effects model is calculated using quadrature. As the panel sizes (or ρ) increase, the quadrature approximation becomes less accurate. We can use the quadchk command to see if changing the number of quadrature points affects the results. If the results do change, then the quadrature approximation is not accurate and the results of the model should not be interpreted. See [XT] **quadchk** for details and [XT] **xtprobit** for an example.

❏

Saved Results

   xttobit saves in e():

Scalars

e(N)	number of observations	e(g_min)	smallest group size
e(N_g)	number of groups	e(g_avg)	average group size
e(N_unc)	number of uncensored observations	e(chi2)	χ^2
e(N_lc)	number of left-censored observations	e(chi2_c)	χ^2 for comparison test
e(N_rc)	number of right-censored observations	e(rho)	ρ
e(df_m)	model degrees of freedom	e(sigma_u)	panel-level standard deviation
e(ll)	log likelihood	e(sigma_e)	standard deviation of ϵ_{it}
e(ll_0)	log likelihood, constant-only model	e(N_cd)	number of completely determined obs.
e(ll_c)	log likelihood, comparison model	e(n_quad)	number of quadrature points
e(g_max)	largest group size		

Macros

e(cmd)	xttobit	e(chi2type)	Wald or LR; type of model χ^2 test
e(depvar)	name(s) of dependent variable(s)	e(chi2_ct)	Wald or LR; type of model χ^2 test
e(title)	title in estimation output		corresponding to e(chi2_c)
e(ivar)	variable denoting groups	e(offset)	offset
e(wtype)	weight type	e(distrib)	Normal; the distribution of the
e(wexp)	weight expression		random effect
e(llopt)	contents of ll(), if specified	e(crittype)	optimization criterion
e(ulopt)	contents of ul(), if specified	e(predict)	program used to implement predict

Matrices

e(b)	coefficient vector	e(V)	variance–covariance matrix of the estimators

Functions

e(sample)	marks estimation sample

Methods and Formulas

   xttobit is implemented as an ado-file.

   Assuming a normal distribution, $N(0, \sigma_\nu^2)$, for the random effects ν_i, we have the joint (unconditional of ν_i) density of the observed data from the ith panel

$$f(y_{i1}^o, \ldots, y_{in_i}^o | \mathbf{x}_{i1}, \ldots, \mathbf{x}_{in_i}) = \int_{-\infty}^{\infty} \frac{e^{-\nu_i^2/2\sigma_\nu^2}}{\sqrt{2\pi}\sigma_\nu} \left\{ \prod_{t=1}^{n_i} F(y_{it}^o, \mathbf{x}_{it}\boldsymbol{\beta} + \nu_i) \right\} d\nu_i$$

where

$$
F(y_{it}^o, \Delta_{it}) = \begin{cases} \left(\sqrt{2\pi}\sigma_\epsilon\right)^{-1} e^{-(y_{it}^o - \Delta_{it})^2/(2\sigma_\epsilon^2)} & \text{if } y_{it}^o \in C \\[2mm] \Phi\left(\frac{y_{it}^o - \Delta_{it}}{\sigma_\epsilon}\right) & \text{if } y_{it}^o \in L \\[2mm] 1 - \Phi\left(\frac{y_{it}^o - \Delta_{it}}{\sigma_\epsilon}\right) & \text{if } y_{it}^o \in R \end{cases}
$$

where C is the set of noncensored observations, L is the set of left-censored observations, R is the set of right-censored observations, and $\Phi()$ is the cumulative normal distribution. We can approximate the integral with M-point Gauss–Hermite quadrature,

$$
\int_{-\infty}^{\infty} e^{-x^2} g(x)dx \approx \sum_{m=1}^{M} w_m^* g(a_m^*)
$$

where the w_m^* denote the quadrature weights and the a_m^* denote the quadrature abscissas. The log-likelihood L, is then calculated using the quadrature

$$
L = \sum_{i=1}^{n} w_i \log\Big\{ f(y_{i1}^o, \ldots, y_{in_i}^o | \mathbf{x}_{i1}, \ldots, \mathbf{x}_{in_i}) \Big\}
$$

$$
\approx \sum_{i=1}^{n} w_i \log\left\{ \frac{1}{\sqrt{\pi}} \sum_{m=1}^{M} w_m^* \prod_{t=1}^{n_i} F\left(y_{it}^o, \mathbf{x}_{it}\boldsymbol{\beta} + \sqrt{2}\sigma_v a_m^*\right) \right\}
$$

where w_i is the user-specified weight for panel i; if no weights are specified, $w_i = 1$.

The quadrature formula requires that the integrated function be well-approximated by a polynomial. As the number of time periods becomes large (as panel size gets large),

$$
\prod_{t=1}^{n_i} F(y_{it}^o, \mathbf{x}_{it}\boldsymbol{\beta} + \nu_i)
$$

is no longer well-approximated by a polynomial. As a general rule of thumb, you should use this quadrature approach only for small to moderate panel sizes (based on simulations, 50 is a reasonably safe upper bound). However, if the data really come from random-effects tobit and rho is not too large (less than, say, .3), then the panel size could be 500 and the quadrature approximation would still be fine. If the data are not random-effects tobit or rho is large (bigger than, say, .7), then the quadrature approximation may be poor for panel sizes larger than 10. The quadchk command should be used to investigate the applicability of the numeric technique used in this command.

References

Honoré, B. 1992. Trimmed LAD and least squares estimation of truncated and censored regression models with fixed effects. *Econometrica* 60: 533–565.

Neuhaus, J. M. 1992. Statistical methods for longitudinal and clustered designs with binary responses. *Statistical Methods in Medical Research* 1: 249–273.

Pendergast, J. F., S. J. Gange, M. A. Newton, M. J. Lindstrom, M. Palta, and M. R. Fisher. 1996. A survey of methods for analyzing clustered binary response data. *International Statistical Review* 64: 89–118.

Also See

Complementary:	[XT] **quadchk**, [XT] **xtdata**, [XT] **xtdes**, [XT] **xtsum**, [XT] **xttab**, [R] **adjust**, [R] **lincom**, [R] **mfx**, [R] **nlcom**, [R] **predict**, [R] **predictnl**, [R] **test**, [R] **testnl**, [R] **vce**
Related:	[XT] **xtgee**, [XT] **xtintreg**, [XT] **xtreg**, [XT] **xtregar**, [R] **tobit**
Background:	[U] **16.5 Accessing coefficients and standard errors**, [U] **23 Estimation and post-estimation commands**, [XT] **xt**

Subject and author index

This is the subject and author index for the *Stata Cross-Sectional Time-Series Reference Manual*. Readers interested in topics other than cross-sectional time-series and graphics, should see the combined subject index at the end of Volume 4 of the *Stata Base Reference Manual*, which indexes the *Stata Base Reference Manual*, the *Stata User's Guide*, the *Stata Programming Reference Manual*, the *Stata Cluster Analysis Reference Manual*, the *Stata Survey Data Reference Manual*, the *Stata Survival Analysis & Epidemiological Tables Reference Manual*, the *Stata Time-Series Reference Manual*, and this manual.

Readers interested in graphics topics should see the index at the end of the *Stata Graphics Reference Manual*.

Semicolons set off the most important entries from the rest. Sometimes no entry will be set off with semicolons; this means all entries are equally important.